Lecture Notes in Physics

Volume 907

The Lecture Notes in Physics

The series Lecture Notes in Physics (LNP), founded in 1969, reports new developments in physics research and teaching-quickly and informally, but with a high quality and the explicit aim to summarize and communicate current knowledge in an accessible way. Books published in this series are conceived as bridging material between advanced graduate textbooks and the forefront of research and to serve three purposes:

- to be a compact and modern up-to-date source of reference on a well-defined topic
- to serve as an accessible introduction to the field to postgraduate students and nonspecialist researchers from related areas
- to be a source of advanced teaching material for specialized seminars, courses and schools

Both monographs and multi-author volumes will be considered for publication. Edited volumes should, however, consist of a very limited number of contributions only. Proceedings will not be considered for LNP.

Volumes published in LNP are disseminated both in print and in electronic formats, the electronic archive being available at springerlink.com. The series content is indexed, abstracted and referenced by many abstracting and information services, bibliographic networks, subscription agencies, library networks, and consortia.

Proposals should be sent to a member of the Editorial Board, or directly to the managing editor at Springer:

Christian Caron
Springer Heidelberg
Physics Editorial Department I
Tiergartenstrasse 17
69121 Heidelberg/Germany
christian.caron@springer.com

More information about this series at http://www.springer.com/series/5304

Valerio Faraoni

Cosmological and Black Hole Apparent Horizons

Springer

Valerio Faraoni
Physics Department
Bishop's University
Sherbrooke, QC, Canada

ISSN 0075-8450   ISSN 1616-6361   (electronic)
Lecture Notes in Physics
ISBN 978-3-319-19239-0   ISBN 978-3-319-19240-6   (eBook)
DOI 10.1007/978-3-319-19240-6

Library of Congress Control Number: 2015943318

Springer Cham Heidelberg New York Dordrecht London

Printed on acid-free paper

Springer International Publishing AG Switzerland is part of Springer Science+Business Media (www.
springer.com)

To my sisters Adriana and Lilly

Preface

Currently, substantial research efforts are devoted to understanding the physics of horizons. A horizon is a surface which separates a region of space-time which is accessible to an observer from one which is not and from which this observer cannot receive light or other physical signals. This feature gives rise to interesting physics: with the emphasis given in modern times to the role played by information in theoretical physics, it is easy to guess that a horizon will produce interesting physical phenomena. If entropy is understood as information entropy, then a horizon which hides information should be attributed some entropy. This is in fact what the black hole thermodynamics developed in the 1970s found. The discovery of Hawking radiation from black hole horizons made possible the development of black hole thermodynamics, a remarkable and beautiful construct which shows that, indeed, there is very interesting physics associated with horizons. Already in special relativity without gravity, uniformly accelerated observers experience acceleration horizons. When gravity is introduced, one encounters black hole and cosmological horizons. Then, studying black holes, one meets inner, outer, Cauchy, and extremal horizons, and in cosmology there are particle, event, de Sitter, and apparent horizons.

The pioneers who developed black hole mechanics and thermodynamics in the 1970s discussed *stationary* black holes and *event* horizons. Dynamical situations such as gravitational collapse, black hole evaporation by Hawking radiation, and black holes interacting with nontrivial environments and exchanging mass-energy require that the concept of event horizon be generalized. Conceivable dynamical situations include black holes accreting surrounding fluids, black holes immersed in a cosmological background, and, most significantly, black holes emitting (and possibly absorbing) Hawking radiation, which becomes significant in the last evolutionary stages. If a black hole is placed in a nontrivial environment, its mass-energy should be also the internal energy which we need to account for in the first law of thermodynamics. This mass-energy must be defined carefully, usually with some quasi-local notion, which in turn is sometimes related to the notion of *apparent* horizon.

In dynamical situations, it is not clear what is meant by "black hole" because the most salient feature of a black hole is precisely its horizon, and the event horizon familiar from stationary black holes turns out to be essentially useless for practical purposes in dynamical space-times. This major obstacle appears because the definition of event horizon requires the knowledge of the entire future of space-time, which is physically impossible to achieve in nonstationary situations. The ambiguity in the notion of "horizon" therefore implies murkiness in the concept of "black hole" itself. Simultaneously, thanks to the increase in the power of modern supercomputers, great theoretical efforts are now made to predict in detail the waveforms of gravitational waves emitted by black holes. These waveforms are needed to build banks of templates to separate signals from noise in the laser interferometric detectors of gravitational waves. The notion of event horizon is of little use in the numerical study of fast astrophysical processes producing those gravitational waves. Instead, "black holes" are routinely identified with outermost marginally trapped surfaces and apparent horizons in numerical research. Hence, a part of the research community is still focused on event horizons, while another part dismisses it altogether and uses horizon surfaces, the role of which is not yet understood clearly. This dichotomy needs to be addressed, and this work is intended to give a contribution in this direction.

This book contains a series of graduate-level lectures introducing the main problems in this area of theoretical physics. The first three chapters are pedagogical in nature, while the remaining two report a series of "case studies" to which the concepts of apparent and trapping horizon are applied. They consist of relatively rare analytic solutions of Einstein's theory and of scalar-tensor and $f(\mathcal{R})$ gravity which appeared in the literature and contain, at least in certain space-time regions, black hole and cosmological apparent horizons. The dynamics of apparent horizons can be rather bizarre and reserves several surprises: The phenomenology of apparent horizons known thus far is described and analyzed. While this field of research is definitely not settled and the last word is not said on any of the issues examined, these lectures aim at collecting and summarizing the existing results and providing an introduction and a toolkit for researchers approaching this field, especially graduate students. An extensive bibliography refers the reader to specific points which cannot be discussed in a single volume. I hope that these lectures will be stimulating and that some of my readers will soon find new directions for this area of research.

Sherbrooke, Canada   Valerio Faraoni
April 2015

Acknowledgments

I am indebted to many people for their contributions to the material discussed in this book, and my thanks go first of all to my collaborators and to my students: Thomas Sotiriou, Vincenzo Vitagliano, Stefano Liberati, Angus Prain, Alex Nielsen, Audrey Jacques, Andres Zambrano, and Roshina Nandra.

For discussions or comments, or for the occasional e-mail pointing out references that I missed, I am grateful to Hideki Maeda, Matt Visser, Daniel Guariento, John Barrow, Eric Poisson, Ivan Booth, Luciano Vanzo, Sergio Zerbini, Rituparno Goswami, Bibhas Majhi, Andrzej Krasiński, Kayll Lake, Viqar Husain, Timothy Clifton, Sergei Odintsov, Emilio Elizalde, Sebastiano Sonego, Nemanja Kaloper, Bahram Mashhoon, Salvatore Capozziello, Valeri Frolov, Sarah Shandera, Carlos Herdeiro, José Senovilla, and Rong-Gen Cai. Many thanks to Angus Prain for drawing the figures and to Aldo Rampioni and Kirsten Theunissen of Springer for their support and friendly assistance during the writing of this book.

This research is supported by Bishop's University and by the Natural Sciences and Engineering Research Council of Canada.

Contents

Symbols and Acronyms

g_{ab}	metric tensor
g^{ab}	inverse metric tensor
g	metric determinant
η_{ab}	Minkowski metric
∇_a	covariant derivative operator
Γ^a_{bc}	Christoffel symbols
R_{ab}	Ricci tensor
$\mathscr{R} = R^a{}_a$	Ricci scalar
$G_{ab} \equiv R_{ab} - \frac{1}{2} g_{ab} R$	Einstein tensor
T_{ab}	matter energy-momentum tensor
Λ	cosmological constant
l_{Pl}	Planck length
k^a, ξ^a	Killing vectors
K^a	Kodama vector
κ	surface gravity
τ	Kodama time
a	uniform acceleration of Rindler observers
$d\Omega^2_{(2)} = d\theta^2 + \sin^2\theta \, d\varphi^2$	line element on the unit 2-sphere
R	areal radius
$\bar{r}$	isotropic radius
η	conformal time
η^a	geodesic deviation vector
θ_l	expansion of a null geodesic congruence with tangent field l^a
M_{MSH}	Misner-Sharp-Hernandez mass
k_B	Boltzmann constant
T	(absolute) temperature
S	entropy
$\mathscr{A}$	horizon area

k	curvature index of FLRW space
a	scale factor of the FLRW metric
$H \equiv \dot{a}/a$	Hubble parameter of FLRW space
$q \equiv -\ddot{a}a/\dot{a}^2$	deceleration parameter of FLRW space
$\dot{f}$	derivative of f with respect to comoving time
$\mathscr{I}^+$	future null infinity
$J^- \left(\mathscr{I}^+ \right)$	causal past of future null infinity
$\equiv$	equal by definition
$\doteq$	equality valid in General Relativity in FLRW space with a perfect fluid as the matter source

FLRW	Friedmann-Lemaître-Robertson-Walker
AH	apparent horizon
TH	trapping horizon
EH	event horizon·
PH	particle horizon
FOTH	future outer trapping horizon
FITH	future inner trapping horizon
PITH	past inner trapping horizon
MOTS	marginally outer trapped surface
MTT	marginally trapped tube
MOTT	marginally outer trapped tube
iff	if and only if

Chapter 1
Stationary Black Holes in General Relativity

The noblest pleasure is the joy of understanding.

—Leonardo da Vinci

1.1 Introduction

Beginning with the Rindler horizons which appear for accelerated observers in Minkowski space without gravity, one comes quickly to acknowledge the presence of various types of horizons when gravity is introduced in spacetime: first one encounters black hole horizons and cosmological horizons. Then, studying classical and semiclassical black holes one is faced with inner, outer, Cauchy, and extremal horizons. The early literature on black holes and the works which developed black hole thermodynamics in the seventies had plenty to do with discussing stationary black holes and *event* horizons (e.g., [23, 74, 75]). Dynamical situations such as gravitational collapse, black hole evaporation due to Hawking radiation, and black holes interacting with non-trivial environments and exchanging mass-energy require that the concept of event horizon be generalized to some other construct with which it is possible to work. Conceivable dynamical situations include black holes accreting or expelling gravitating (i.e., non-test) matter; examples are Vaidya spacetimes, black holes immersed in a cosmological "background" other than de Sitter space, black holes emitting (and possibly also absorbing) Hawking radiation (which becomes significant in the last evolutionary stages with backreaction playing an important role), or black holes with variable mass because of other physical processes. If a black hole is placed in a non-trivial environment, its mass-energy should be also the internal energy which we need to account for in the first law of thermodynamics. This mass-energy must be defined carefully; usually it is identified with some quasi-local energy construct which is related to the notion of horizon. In these lectures we will use the Misner-Sharp-Hernandez mass in spherical symmetry and its generalization, the Hawking-Hayward quasi-local energy in the absence of spherical symmetry.

Intuitively, an horizon is "a frontier between things observable and things unobservable" [64]. Inequivalent notions of black hole horizon abound in the technical literature and the terminology used features event, Killing, inner, outer, Cauchy, apparent, trapping, quasi-local, isolated, dynamical, and slowly evolving

© Springer International Publishing Switzerland 2015
V. Faraoni, *Cosmological and Black Hole Apparent Horizons*,
Lecture Notes in Physics 907, DOI 10.1007/978-3-319-19240-6_1

horizons (Refs. [4, 7, 51, 74] are reviews of an extensive literature). For stationary black holes some of these constructs coincide, but they are, in general, very different or unrelated for dynamical black holes with masses and other physical parameters which change with time. It is not clear what it is meant by "black hole" in dynamical situations because the most salient feature of a black hole is precisely its horizon, which is universally taken to signal the presence of a black hole. The ambiguity in the appropriate notion of "horizon" therefore implies a serious ambiguity in the concept of "black hole".

The definition of event horizon inspired by (and historically attached to) stationary black holes turns out to be essentially useless for practical purposes in dynamical spacetimes. This major obstacle manifests itself because knowing the event horizon requires the knowledge of future null infinity, which is physically impossible to achieve [3].

Astronomy is undergoing remarkable progress and it points out more and more the important roles of stellar mass and supermassive black holes in astrophysical processes. Great theoretical efforts are made to predict in detail the waveforms of gravitational waves emitted by black holes. This programme is made possible by the increase in power of modern supercomputers but it remains a very ambitious goal. The notion of event horizon is of little use in the numerical study of the fast dynamical evolution occurring in the gravitational collapse of a cosmic body, or in the close inspiralling and merger of a black hole with its companion in a binary system. "Black holes" are routinely identified with outermost marginally trapped surfaces and apparent horizons in numerical work [6, 13, 73].

What about cosmological horizons? These surfaces are probably the playground in which one should take baby steps in understanding horizon physics. The cosmology textbooks discuss particle and event horizons in relation with early universe inflation [42, 46, 49]. Different cosmological horizons, the apparent and trapping horizons, are also used more and more. It was not long after the discovery of Hawking radiation [32, 34] and the completion of black hole thermodynamics that Gibbons and Hawking pointed out [28] that the event horizon of de Sitter space behaves as a thermodynamic system and should be endowed with a temperature and an entropy. The region of de Sitter space below the de Sitter horizon is static and the horizon itself does not change in time, so it can be regarded to a certain extent as the cosmological analogue of the Schwarzschild event horizon. There is an important difference, though: a de Sitter horizon depends on the observer while the Schwarzschild horizon does not. If the analogy carries through, then the analogue of time-dependent black hole horizons would necessarily be the apparent and trapping horizons of Friedmann-Lemaître-Robertson-Walker (FLRW) spacetime, which evolve with the cosmic time.

Similar to black hole thermodynamics for event horizons, there have been many attempts to formulate a meaningful thermodynamics for other horizon constructs (e.g., [4, 14, 35]). The thermodynamics of black hole apparent, trapping, isolated, and dynamical horizons has been scrutinized often in recent years and thermodynamical studies of FLRW apparent horizons have also appeared.

At the same time, there has been a resurgence of interest in theories of gravity alternative to General Relativity: they are motivated by various reasons. First, there is the search for a quantum theory of gravity, which promotes interest in low-energy effective actions, which invariably contain ingredients foreign to Einstein gravity, such as scalar fields coupled non-minimally to the curvature (which give a scalar-tensor nature to the theory), higher derivative terms, or perhaps non-local terms. Other sources of interest in alternative gravity are the serious attempts [17, 68] to explain the current acceleration of the universe discovered with type Ia supernovae without invoking an ad hoc dark energy [1]. Although not as well motivated, there are also attempts to remove the need for dark matter in galaxies and clusters by modifying gravity, given that dark matter particles still elude direct detection.

Black holes in alternative theories of gravity evade the no-hair theorem of General Relativity and one can have non-trivial scalar hair due to interactions between scalar fields and astrophysical black holes [15, 63, 69, 70], which could lead to detectable effects [5, 36]. Analytic solutions describing time-dependent black holes in these theories allow one to get a glimpse of the phenomenology to expect.

More interest in alternative gravity comes from the thermodynamics of spacetime idea, according to which Einstein theory corresponds somehow to a state of thermodynamic equilibrium [38] and extended gravities to some sort of excitation [19] in some "space of theories". This idea, which seems to fit well in the wider context of emergent gravity approaches to the problem of quantizing gravity, has its foundations in the analysis of local Rindler horizons of observers with worldlines threading spacetime, and in the prescription that the entropy is equal to one quarter of the horizon area.

There are many instances in which cosmological and time-evolving horizons play a role in theoretical research in gravity. In these lectures we review the main properties of cosmological and black hole time-varying horizons in both General Relativity and extended theories of gravity, and we attempt to provide a unified view of their physics for applications to various areas of gravitational theory. We begin with cosmological horizons, and then we discuss horizons associated with time-dependent black holes. Since only a few exact solutions of General Relativity and of other theories of gravity are known for which the horizons are explicitly time-dependent, we concentrate on spacetimes which describe black holes embedded in cosmological "backgrounds", which have been studied in some detail.

We begin our study by reviewing basic material in the first two chapters. First we recall the stationary black hole solutions of the Einstein equations: this first chapter is meant to be only a refresher since there is no point in repeating the vast and excellent literature on "standard" black holes. Chapter 2 reminds the reader of the basic tools used in the analysis of black hole and other horizons, i.e., null geodesic congruences, and introduces various definitions of horizons appearing in the literature. The following chapter analyzes the various horizons of FLRW space, including the thermodynamics proposed for these horizons. We also discuss several coordinate systems for FLRW spaces which are useful in the study of the dynamics and thermodynamics of cosmological horizons. The following chapters

discuss analytic solutions of the field equations of various theories of gravity which
exhibit time-varying horizons and, often, horizons which appear and/or disappear in
pairs.

In the following, a spacetime $(\mathcal{M}, g_{ab})$ is described by a 4-dimensional manifold
$\mathcal{M}$ on which a metric tensor field g_{ab} with Lorentzian signature $-+++$ is defined.
We follow the notations and conventions of Wald's book [74]. In particular, we use
units in which the speed of light c and Newton's constant G are unity. The Riemann
tensor is given by

$$R_{abc}{}^{d} = \Gamma_{ac,b}^{d} - \Gamma_{bc,a}^{d} + \Gamma_{ac}^{e}\Gamma_{eb}^{d} - \Gamma_{bc}^{e}\Gamma_{ea}^{d} \tag{1.1}$$

in terms of the Christoffel symbols Γ_{ab}^{c} of the metric g_{ab}. The Ricci tensor is the
contraction

$$R_{ac} = R_{abc}{}^{b}, \tag{1.2}$$

and the Ricci scalar is $\mathscr{R} = g^{ab}R_{ab}$.

1.2 Stationary Black Holes of General Relativity

For introductions to the theory of General Relativity and basic properties of its
stationary black hole solutions, we refer the reader to well known textbooks
[11, 48, 58, 74]; for a useful list of references on "standard" general-relativistic black
holes see [16, 24, 58]. Here we review background material used in the following
chapters and the main asymptotically flat black hole solutions of Einstein theory.
According to the *no-hair theorems* [12, 33, 37, 50, 66, 67], the Schwarzschild and
Kerr spacetimes and their charged (Reissner-Nordström and Kerr-Newman) generalizations are the generic asymptotically flat electrovacuum black hole solutions
of this theory. In General Relativity a black hole formed by gravitational collapse
will settle down to a state determined by only three parameters: its mass M, angular
momentum J, and electric charge Q, irrespective of the initial configuration, the
nature of the collapsing matter, and the details of the collapse. Perturbation analyses
show that perturbations are radiated away quickly according to laws established by
Price (a field with spin s will radiate away a multipole $l \geq s$ in such a manner
that, in the late stage of collapse, the field decays with a power-law tail scaling with
time as t^{2l+p+1}, where $p = 1$ for initially static multipoles and $p = 2$ otherwise
[30, 60, 61]). A black hole characterized only by the three parameters M, J, and Q
can correspond to a very large number of possible configurations unobservable by
an observer located outside the horizon and then, heuristically, to a large entropy.

Let us review the classic black hole spacetimes in various coordinate systems. It
is often convenient to introduce coordinates tied to particular families of timelike
observers or to use null coordinates based on outgoing or ingoing null geodesics. It

is sometimes useful to see the latter as the null limit of the worldlines of timelike observers. All the spacetimes which we review in this chapter are asymptotically flat and stationary.

1.3 Schwarzschild Spacetime

The Schwarzschild metric was discovered soon after Einstein introduced the theory of General Relativity and is the prototype of a black hole spacetime. Several coordinate systems have been developed in order to study it.

1.3.1 Schwarzschild Coordinates

The *Schwarzschild line element* is

$$ds^2 = -\left(1 - \frac{2M}{R}\right) dt^2 + \frac{dR^2}{1 - \frac{2M}{R}} + R^2 d\Omega_{(2)}^2 \qquad (1.3)$$

in Schwarzschild (or "curvature") coordinates in which R is the areal radius, i.e., $R \equiv \sqrt{\frac{\mathscr{A}}{4\pi}}$, with $\mathscr{A}$ being the area of 2-spheres of symmetry, and where

$$d\Omega_{(2)}^2 \equiv d\theta^2 + \sin^2 \theta \, d\varphi^2 \qquad (1.4)$$

is the line element on the unit 2-sphere, which will be used throughout these lectures. The metric (1.3) is a vacuum solution of the Einstein equations and it is static, spherically symmetric, and asymptotically flat; it exhibits the well-known event horizon at $R = 2M$, which corresponds to a singularity of the Schwarzschild coordinates at which $g_{00} = 0$, but not to a true spacetime singularity—the invariants of the curvature tensor are finite there. There is a true spacetime singularity at $R = 0$, where the invariants of the curvature tensor diverge. The Schwarzschild coordinate patch only covers the region $R > 2M$ exterior to the horizon.

The concept of event horizon will be discussed in detail in Sect. 2.4. For now, it is sufficient to know that light and massive particles which start out in the region $R < 2M$ cannot escape from it. The entire region $R < 2M$ "below the horizon" is not seen by observers located at radii $R > 2M$ (the region "outside the horizon").

The Schwarzschild line element represents a one-parameter family of metrics parametrized by the mass parameter M. Only non-negative values of this parameter are physical, while the limit $M \to 0$ gives Minkowski space.[1]

---

[1]One should be careful, however, in taking limits of the spacetime geometry based on coordinate systems: the $M \to 0$ limit of Schwarzschild space, really, produces either Minkowski space or a

1.3.2   Isotropic Coordinates

It is common to see the Schwarzschild metric presented using the *isotropic radius* $\bar{r}$ defined by

$$R \equiv \bar{r}\left(1 + \frac{M}{2\bar{r}}\right)^2. \tag{1.5}$$

Isotropic coordinates $(t, \bar{r}, \theta, \varphi)$ consist of the usual Schwarzschild time t and polar coordinates (θ, φ), with the Schwarzschild areal radius R replaced by the isotropic radius $\bar{r}$. Using $\bar{r}$, there are two copies of the spacetime region $R > 2M$ outside the horizon because Eq. (1.5) gives

$$\bar{r} = \frac{R}{2}\left(1 - \frac{M}{R} \pm \sqrt{1 - \frac{2M}{R}}\right). \tag{1.6}$$

Isotropic coordinates for the Schwarzschild geometry were studied in detail in Refs. [8, 76].

The definition (1.5) gives the relation between differentials

$$dR = \left(1 - \frac{M^2}{4\bar{r}^2}\right)d\bar{r} \tag{1.7}$$

and, using the relation

$$1 - \frac{2M}{R} = \left(\frac{1 - \frac{M}{2\bar{r}}}{1 + \frac{M}{2\bar{r}}}\right)^2, \tag{1.8}$$

it is easy to obtain the *Schwarzschild line element in isotropic coordinates* from Eq. (1.3)

$$ds^2 = -\frac{\left(1 - \frac{M}{2\bar{r}}\right)^2}{\left(1 + \frac{M}{2\bar{r}}\right)^2}dt^2 + \left(1 + \frac{M}{2\bar{r}}\right)^4\left(d\bar{r}^2 + \bar{r}^2 d\Omega_{(2)}^2\right). \tag{1.9}$$

The Schwarzschild event horizon $R = 2M$ corresponds to $\bar{r} = M/2$, where the two values (1.6) of the coordinate $\bar{r}$ coincide.

Kasner metric [27] and, strictly speaking, a coordinate-free approach [55–57] is needed to make limits rigorous.

1.3.3 Kruskal-Szekeres Coordinates

The Kruskal-Szekeres coordinates [44, 72] replace the Schwarzschild coordinates (t, R) leaving the polar angles (θ, φ) untouched and are based on ingoing and outgoing radial null geodesics. Introduce first the Regge-Wheeler tortoise coordinate [62, 77]

$$r^* \equiv R + 2M \ln \left| \frac{R}{2M} - 1 \right| \tag{1.10}$$

defined in the range $-\infty < r^* < +\infty$ corresponding to $2M < R < +\infty$. This coordinate is chosen so that

$$\frac{dR^2}{1 - 2M/R} = \left(1 - \frac{2M}{R}\right)(dr^*)^2 \tag{1.11}$$

and the 2-metric of the (t, r^*) surface is explicitly conformally flat,

$$ds_{(2)}^2 = \left(1 - \frac{2M}{R}\right)\left[-dt^2 + (dr^*)^2\right]. \tag{1.12}$$

The differential of r^* is related to that of the areal radius R by

$$dr^* = \frac{dR}{1 - 2M/R}. \tag{1.13}$$

Then the null coordinates

$$u \equiv t - r^*, \qquad v \equiv t + r^* \tag{1.14}$$

("retarded time" and "advanced time", respectively) turn the line element (1.3) into

$$ds^2 = \left(1 - \frac{2M}{R}\right)\left[-dt^2 + (dr^*)^2\right] + R^2 d\Omega_{(2)}^2 = -\left(1 - \frac{2M}{R}\right) du\, dv + R^2 d\Omega_{(2)}^2, \tag{1.15}$$

where $R(u, v)$ is an implicit function of u and v defined by $r^*(R) = (v - u)/2$. Now introduce the new null coordinates

$$U \equiv \mp e^{-u/4M}, \qquad V \equiv e^{v/4M}, \tag{1.16}$$

where the upper sign refers to the exterior region (there are two copies of the region $R > 2M$ in these coordinates). The function $R(U, V)$ is given implicitly by

$$e^{R/2M}\left(\frac{R}{2M} - 1\right) = -UV \tag{1.17}$$

and one obtains the *Schwarzschild line element in Kruskal-Szekeres coordinates*

$$ds^2 = -\frac{32M^3}{R}\,e^{-R/2M}\,dUdV + R^2 d\Omega_{(2)}^2 \tag{1.18}$$

which is clearly regular at $R = 2M$.

The surfaces $R = $ constant correspond to $UV = $ constant, which describes hyperbolae with two branches asymptotic to the U and V axes. The event horizon $R = 2M$ corresponds to $U = 0$ and/or $V = 0$, while the singularity $R = 0$ corresponds to two branches of the corresponding hyperbola $UV = 1$. The spacetime region $R > 2M$ covered by the Schwarzschild coordinates corresponds to $V > 0$ and $U < 0$ ("region I" in Fig. 1.1), but it is clear that the Schwarzschild manifold includes also the region $U > 0$ and $V > 0$ ("region II"). There are also two other regions ("region III" and "region IV") describing a Schwarzschild white hole, i.e., the time-reversal of a black hole, which constitutes the maximal extension of the Schwarzschild manifold. However, these regions are not accessible to timelike or null particles because the Kruskal diagram describes vacuum and must be cut off at the timelike surface of collapsing matter. The $R = 0$ singularity of the Schwarzschild metric is *spacelike* because the coordinate R turns timelike when crossing from outside to inside the event horizon $R = 2M$ and for $0 < R < 2M$ a surface $R = $ constant is spacelike (much like a surface $t = $ constant in the region $R > 2M$ outside the event horizon).

1.3.4   Eddington-Finkelstein Coordinates

The Eddington-Finkelstein coordinates [18, 21] use either the retarded or the advanced time (1.14). The *ingoing* Eddington-Finkelstein coordinates are (v, R, θ, φ), the *outgoing* Eddington-Finkelstein coordinates are (u, R, θ, φ).

In *ingoing Eddington-Finkelstein coordinates* (v, R, θ, φ), using Eq. (1.13) and $dt = dv - dr^* = dv - dR/(1 - 2M/R)$, one obtains

$$ds^2 = -\left(1 - \frac{2M}{R}\right) dv^2 + 2dvdR + R^2 d\Omega_{(2)}^2. \tag{1.19}$$

In *outgoing Eddington-Finkelstein coordinates* (u, R, θ, φ), $dt = du + dr^* = du + \frac{dR}{1-2M/R}$ gives

$$ds^2 = -\left(1 - \frac{2M}{R}\right) du^2 - 2dudR + R^2 d\Omega_{(2)}^2. \tag{1.20}$$

(Note the difference in the sign of the off-diagonal term in outgoing and ingoing coordinates.) The coordinates (v, R) cover regions I and II of the Kruskal-Szekeres diagram, while the coordinates (u, R) cover regions III and IV. Clearly, since u

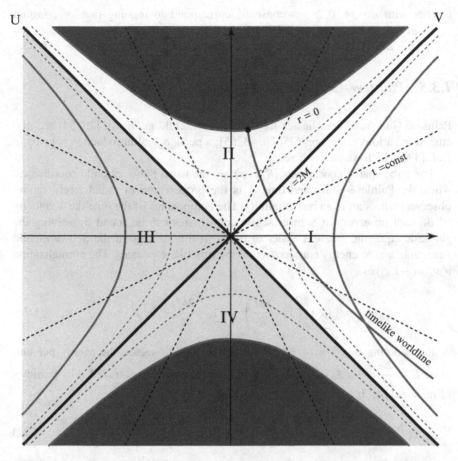

Fig. 1.1 The Kruskal-Szekeres plane: the horizon $R = 2M$ corresponds to the U and V axes, the singularity $R = 0$ to the hyperbola $UV = 1$, and region I is covered by the Schwarzschild coordinates (t, R). The Schwarzschild black hole is described by regions I and II and the white hole by regions III and IV. The timelike worldline of a particle crossing the event horizon and falling onto the singularity is also shown

describes outward-propagating radial null rays and these cannot exit from the Schwarzschild event horizon, the coordinates (u, R) cannot describe regions I and II but they are useful to describe the white hole regions. In an Eddington-Finkelstein (v, R) or (u, R) diagram, outgoing radial null geodesics do not propagate at 45° angles while ingoing radial null geodesics do. In fact radial null geodesics, which have $ds^2 = 0$ and $d\theta = d\varphi = 0$, satisfy

$$dv\left[-\left(1 - \frac{2M}{R}\right)dv + 2dR\right] = 0.$$

Curves with $dv = 0$ ($v = $ constant) correspond to ingoing rays of constant slope $dv/dR = 0$ at 45° angles with the axes of the (t, R) plane, while $dv = 2dR/(1 - 2M/R)$ corresponds to rays of variable slope $dv/dR = 2/(1 - 2M/R)$.

1.3.5 Painlevé-Gullstrand Coordinates

Painlevé-Gullstrand coordinates for the Schwarzschild geometry [29, 54] are discussed in various references [25, 26, 43, 65]; a pedagogical introduction is given in Ref. [47], which we follow here.

The three spatial coordinates (R, θ, φ) are the usual Schwarzschild coordinates, while the Painlevé-Gullstrand time T is the proper time of radial freely falling observers who start from rest at infinity. The components of the timelike 4-velocity u^a of such observers in Schwarzschild coordinates can be found by solving the geodesic equation, but it is easier to resort to the staticity of the Schwarzschild spacetime and to energy conservation along timelike geodesics. The normalization $u^c u_c = -1$ gives

$$(u^0)^2 \left(1 - \frac{2M}{R}\right)^2 = 1 - \frac{2M}{R} + (u^1)^2 \tag{1.21}$$

while, denoting by $\xi^a = (\partial/\partial t)^a$ the timelike Killing vector, the energy per unit mass $\tilde{E} \equiv \dfrac{E}{m} = -u^c \xi_c$ is conserved along timelike geodesics travelled by particles of mass m [74]. It is

$$\tilde{E} = -u^c \xi_c = -u^\mu \cdot (-) \left(1 - \frac{2M}{R}\right) \delta_{\mu 0} = \left(1 - \frac{2M}{R}\right) u^0 \tag{1.22}$$

which, in conjunction with Eq. (1.21), yields

$$\tilde{E}^2 = 1 - \frac{2M}{R} + (u^1)^2 . \tag{1.23}$$

Since the observer is infalling, it is $u^1 < 0$. At $R \to +\infty$, this relation yields

$$\tilde{E}^2 = 1 + \left(\frac{v_{(\infty)}}{\sqrt{1 - v_{(\infty)}^2}}\right)^2 = \frac{1}{1 - v_{(\infty)}^2} \equiv \gamma_{(\infty)}^2 ,$$

where $v_{(\infty)}$ is the three-dimensional velocity at infinity and $\gamma_{(\infty)}$ is the corresponding Lorentz factor. The energy per unit mass of this observer can be expressed as

$$\tilde{E} = u^0_{(\infty)} = \frac{1}{\sqrt{1 - v_{(\infty)}^2}} . \tag{1.24}$$

The Painlevé-Gullstrand time T is obtained by setting[2] $v_{(\infty)} = 0$. With this choice one obtains $\tilde{E} = 1$, $u^0 = (1 - 2M/R)^{-1}$, and $u^1 = -\sqrt{\dfrac{2M}{R}}$. Therefore, one has

$$u^\mu = \left(\frac{1}{1 - \frac{2M}{R}}, -\sqrt{\frac{2M}{R}}, 0, 0 \right),$$ (1.25)

$$u_\mu = \left(-1, -\frac{\sqrt{2M/R}}{1 - \frac{2M}{R}}, 0, 0 \right).$$ (1.26)

Now, it is

$$u_\mu = -\nabla_\mu T = -\partial_\mu T$$ (1.27)

for a function T; this property is crucial for the introduction of the Painlevé-Gullstrand time T. In fact, by integrating it using Eq. (1.26) one obtains $\partial_t T = 1$ and

$$T = t + f(R)$$ (1.28)

for $\mu = 0$ while, for $\mu = 1$, one obtains $\partial_R T = \dfrac{\sqrt{2M/R}}{1 - 2M/R}$ and

$$T = \int \frac{\sqrt{2M/R}}{1 - 2M/R} dR + g(t).$$ (1.29)

By comparing Eqs. (1.28) and (1.29) one obtains

$$T = t + \int \frac{\sqrt{2M/R}}{1 - 2M/R} dR,$$ (1.30)

which is immediately integrated to

$$T = t + 4M \left(\sqrt{\frac{R}{2M}} + \frac{1}{2} \ln \left| \frac{\sqrt{\frac{R}{2M}} - 1}{\sqrt{\frac{R}{2M}} + 1} \right| \right).$$ (1.31)

---

[2]A more general family of coordinates parametrized by the parameter $p \equiv 1 - v_{(\infty)}^2$ with $0 < v_{(\infty)} < 1$ are introduced in [45] and discussed in [47]; it includes as special cases the Painlevé-Gullstrand coordinates for $p \to 1$, Eddington-Finkelstein coordinates in the lightlike limit $p \to 0$, and it is related to another family of coordinate systems characterized by $p > 1$ and discussed in Refs. [25, 26].

As a check, by differentiating the last expression it is easy to see that T satisfies
Eq. (1.27). Contrary to the Kruskal-Szekeres coordinates, the Painlevé-Gullstrand
time is given as an explicit function of t and R.

Since, from Eq. (1.30), it is $dt = dT - \sqrt{2M/R}\,(1 - 2M/R)^{-1}\,dR$, one obtains
the *Schwarzschild line element in Painlevé-Gullstrand coordinates*

$$ds^2 = -\left(1 - \frac{2M}{R}\right)dT^2 + 2\sqrt{\frac{2M}{R}}\,dTdR + dR^2 + R^2 d\Omega^2_{(2)} \tag{1.32}$$

or, alternatively,

$$ds^2 = -dT^2 + \left(dR + \sqrt{\frac{2M}{R}}\,dT\right)^2 + R^2 d\Omega^2_{(2)}. \tag{1.33}$$

The metric (1.32) is clearly regular at the event horizon $R = 2M$ but singular at
$R = 0$; it is non-diagonal and the three-dimensional surfaces $T = $ constant are flat,
as can be seen by setting $T = $ constant which gives $ds^2_{(3)} = dR^2 + R^2 d\Omega^2_{(2)}$, the
Euclidean line element in three dimensions.

The Painlevé-Gullstrand coordinates do not cover the white hole portion of the
Kruskal-Szekeres plane but only regions I and II because the radial freely falling
observers cross the future, but not the past, event horizon (see Ref. [47] for a
discussion).

Following Ref. [52] (the authors of which actually consider the more complicated
situation of non-static and non-asymptotically flat metrics), we can reintroduce
the speed of light c and define the quantity $v(R) \equiv \sqrt{2M/R}$ to rewrite the line
element (1.32) as

$$ds^2 = -\left[c^2 - v^2(R)\right]dT^2 + 2v(R)dTdR + dR^2 + R^2 d\Omega^2_{(2)}. \tag{1.34}$$

1.3.6   Kerr-Schild Coordinates

A *Kerr-Schild metric* is an algebraically special metric of the form [40, 41, 71]

$$g_{ab} = \eta_{ab} + \lambda\,k_a k_b, \tag{1.35}$$

where η_{ab} is the flat Minkowski metric, λ is a scalar function, and k^a is a null
geodesic vector with respect to both η_{ab} and g_{ab}:

$$\eta_{ab}k^a k^b = 0, \qquad g_{ab}k^a k^b = 0, \tag{1.36}$$

$$k^c \partial_c k^a = k^c \nabla_c k^a = 0, \tag{1.37}$$

where $k^a \equiv \eta^{ab} k_b$. The inverse of the Kerr-Schild metric (1.35) is

$$g^{ab} = \eta^{ab} - \lambda k^a k^b \qquad (1.38)$$

since $g_{ab} g^{bc} = \delta_a^c$. Kerr-Schild coordinates are those in which a Kerr-Schild metric assumes explicitly the form (1.35). The Einstein field equations for Kerr-Schild metrics in Kerr-Schild coordinates are linear [31]. The Kerr-Schild class of metrics includes the Reissner-Nordström, Kerr-Newman, Vaidya, and *pp*-wave spacetimes. Kerr-Schild coordinates for the Schwarzschild metric are $(\tilde{t}, R, \theta, \varphi)$, where the time coordinate used is [48]

$$\tilde{t} = v - R = t + r^* - R. \qquad (1.39)$$

The null geodesic vector corresponds to the tangent of the *ingoing* radial null congruence and the *Schwarzschild metric in Kerr-Schild coordinates* is [48]

$$ds^2 = -\left(1 - \frac{2M}{R}\right) d\tilde{t}^2 + \frac{4M}{R} d\tilde{t}\, dR + \left(1 + \frac{2M}{R}\right) dR^2 + R^2 d\Omega_{(2)}^2, \qquad (1.40)$$

where $\lambda = 2M/R$ and $k_\mu = (1, 1, 0, 0)$ in coordinates $(\tilde{t}, R, \theta, \varphi)$.

1.3.7 Novikov Coordinates

The Novikov coordinates [48, 53] employ the comoving time τ of geodesic observers and the comoving radius

$$R_* \equiv \sqrt{\frac{R_{\max}}{2M} - 1}, \qquad (1.41)$$

where $R_{\max}$ is the largest R-coordinate attained by a test particle ejected near the singularity $R = 0$ [48, 53]. The *Schwarzschild line element in Novikov coordinates* is

$$ds^2 = -d\tau^2 + \left(\frac{1 + R_*^2}{R_*^2}\right)\left(\frac{\partial R}{\partial R_*}\right)^2 dR_*^2 + R^2 d\Omega_{(2)}^2, \qquad (1.42)$$

where $R = R(\tau, R_*)$ is given implicitly by

$$\frac{\tau}{2M} = \pm \left(1 + R_*^2\right) \sqrt{\frac{R}{2M} - \frac{(R/2M)^2}{1 + R_*^2}} + \left(1 + R_*^2\right)^{3/2} \cos^{-1}\left(\sqrt{\frac{R/2M}{1 + R_*^2}}\right). \qquad (1.43)$$

1.4   Reissner-Nordström Metric

The Reissner-Nordström spacetime describes the geometry of a static, spherically symmetric, asymptotically flat, electrically charged black hole which solves the Einstein-Maxwell equations (therefore, not the vacuum, but the "electrovacuum" Einstein equations) with a purely electric radial field.

1.4.1   Schwarzschild Coordinates

The *Reissner-Nordström line element in Schwarzschild coordinates* is

$$ds^2 = -\left(1 - \frac{2m}{R} + \frac{Q^2}{R^2}\right) dt^2 + \left(1 - \frac{2m}{R} + \frac{Q^2}{R^2}\right)^{-1} dR^2 + R^2 d\Omega_{(2)}^2 , \qquad (1.44)$$

while the only non-vanishing components of the Maxwell tensor are those of the electric field

$$F^{01} = -F^{10} = \frac{Q}{R^2} . \qquad (1.45)$$

This line element describes a two-parameter family of metrics parametrized by the mass m (which is the Arnowitt-Deser-Misner mass [2]) and the electric charge Q.

The inverse metric component $g^{11} = \left(1 - \frac{2m}{R} + \frac{Q^2}{R^2}\right)$ vanishes at

$$R_\pm = m \pm \sqrt{m^2 - Q^2} , \qquad (1.46)$$

and therefore there are two horizons, commonly called *inner* and *outer* horizon.[3] The metric (1.44) describes a black hole spacetime when $|Q| \leq m$ and a naked singularity when $|Q| > m$. The case $|Q| = m$ describes an *extremal black hole* for which the inner and outer horizons coincide, $R_+ = R_- = m$.

Since g_{00} vanishes at R_+, the Schwarzschild coordinates (t, R, θ, φ) become singular there and they only cover the region $R > R_+$.

1.4.2   Kruskal-Szekeres Coordinates

A Kruskal-Szekeres coordinate patch can be introduced which covers the region $R_- < R < +\infty$, but it does not penetrate the inner horizon $R = R_-$ and another

---

[3]In the terminology to be introduced later, the outer horizon $R = R_+$ is an event and an apparent horizon, while the inner horizon $R = R_-$ is an apparent, but not an event, horizon, and is also a Cauchy horizon which is unstable [9, 59].

coordinate patch is needed there (that is, Kruskal-Szekeres coordinates are specific to a single horizon). We introduce $u \equiv t - r^*$ and $v \equiv t + r^*$ as usual, the function $f(R) \equiv 1 - \dfrac{2m}{R} + \dfrac{Q^2}{R^2}$, and the quantity

$$\kappa_+ \equiv \frac{f'(R_+)}{2} = \frac{1}{R_+^2}\left(m - \frac{Q^2}{R_+}\right). \tag{1.47}$$

Then, for $R_- < R < +\infty$, we define the Kruskal-Szekeres coordinates

$$U_+ = \mp e^{-\kappa + u}, \qquad V_+ = e^{\kappa + v}, \tag{1.48}$$

with the upper sign for the exterior region $R > R_+$ and the lower one for the interior region $R < R_+$. These coordinates are well behaved near the outer horizon but become singular near the inner horizon where $r^* \to +\infty$ (see Ref. [58] for a detailed discussion). Near the outer horizon the metric in Kruskal-Szekeres coordinates is

$$ds^2 \simeq -\frac{2}{\kappa_+^2}\,dU_+ dV_+ + R_+^2\,d\Omega_{(2)}^2. \tag{1.49}$$

A second patch of Schwarzschild coordinates, distinct from the one used for $R > R_+$, can be used in the region $R_- < R < R_+$. To extend the metric inside the inner horizon, define again $u \equiv t - r^*$ and $v \equiv t + r^*$ using the inner Schwarzschild radial and the tortoise coordinates, in addition to

$$\kappa_- \equiv \frac{f'(R_-)}{2}, \tag{1.50}$$

and then

$$U_- = \mp e^{-\kappa - u}, \qquad V_- = -e^{\kappa - v}, \tag{1.51}$$

with the upper sign for $R > R_-$ and the lower one for $R < R_-$. Near the inner horizon $R = R_-$ the line element is [58]

$$ds^2 \simeq -\frac{2}{\kappa_-^2}\,dU_- dV_- + R_-^2\,d\Omega_{(2)}^2, \tag{1.52}$$

which is regular at $R = R_-$. The coordinates $(U_-, V_-, 0, 0)$ are regular at the inner horizon but singular at the outer one $R = R_+$.

For $R < R_-$, it is $g_{00} = -f < 0$, hence the singularity at $R = 0$ is *timelike*, contrary to the Schwarzschild $R = 0$ singularity which is spacelike. This means that the Reissner-Nordström singularity can be avoided by observers in the region $R < R_-$, which can go around it in 3-space, while in the Schwarzschild spacetime all observers who have crossed the horizon $R = 2M$ must meet the $R = 0$ singularity within a finite time.

1.5 Kerr Spacetime

The stationary and axially symmetric Kerr metric [39] is interpreted as describing a spinning black hole. The *Kerr line element in Boyer-Linquist coordinates* (t, r, θ, φ) is

$$ds^2 = -\left(1 - \frac{2mr}{\rho^2}\right) dt^2 - \frac{4mar \sin^2 \theta}{\rho^2} dt d\varphi + \frac{\Sigma}{\rho^2} \sin^2 \theta d\varphi^2 + \frac{\rho^2}{\Delta} dr^2 + \rho^2 d\theta^2$$

$$(1.53)$$

or, equivalently,

$$ds^2 = -\frac{\rho^2 \Delta}{\Sigma} dt^2 + \frac{\Sigma}{\rho^2} \sin^2 \theta \, (d\varphi - \omega dt)^2 + \frac{\rho^2}{\Delta} dr^2 + \rho^2 d\theta^2 ,$$

$$(1.54)$$

where

$$\rho^2 = r^2 + a^2 \cos^2 \theta ,$$

$$(1.55)$$

$$\Delta = r^2 - 2mr + a^2 ,$$

$$(1.56)$$

$$\Sigma = \left(r^2 + a^2\right)^2 - a^2 \Delta \sin^2 \theta ,$$

$$(1.57)$$

$$\omega = -\frac{g_{03}}{g_{33}} = \frac{2mar}{\Sigma} .$$

$$(1.58)$$

This line element describes a two-parameter family of spacetimes parametrized by the mass M and the angular momentum per unit mass a (the total angular momentum at spatial infinity given by the Komar formula is $J = Ma$). When $a \to 0$ the Kerr metric reduces to the Schwarzschild one.

The inverse metric has components

$$g^{00} = -\frac{\Sigma}{\rho^2 \Delta} ,$$

$$(1.59)$$

$$g^{03} = g^{30} = -\frac{2Mar}{\rho^2 \Delta} ,$$

$$(1.60)$$

$$g^{11} = \frac{\Delta}{\rho^2} ,$$

$$(1.61)$$

$$g^{22} = \frac{1}{\rho^2} ,$$

$$(1.62)$$

$$g^{33} = \frac{\Delta - a^2 \sin^2 \theta}{\rho^2 \Delta \sin^2 \theta} .$$

$$(1.63)$$

The metric is singular at $\Delta = 0$ and $\rho = 0$; the first surface corresponds to a coordinate singularity while the second one corresponds to a true spacetime singularity. In fact, the Kretschmann scalar

$$R_{abcd}R^{abcd} = \frac{48m^2 \left(r^2 - a^2 \cos^2\theta\right)\left(\rho^4 - 16a^2r^2\cos^2\theta\right)}{\rho^{12}} \tag{1.64}$$

diverges there.

The *static limit* surface is defined by $g_{00} = 0$, which yields

$$r_{\text{SL}} = m + \sqrt{m^2 - a^2\cos^2\theta}. \tag{1.65}$$

This static limit surface is also defined by considering static observers, i.e., observers whose 4-velocity is parallel to the timelike Killing vector $\xi^a = (\partial/\partial t)^a$ of components $\xi^\mu = (1, 0, 0, 0)$,

$$u^a = \gamma\xi^a \equiv \frac{\xi^a}{\sqrt{-g_{cd}\xi^c\xi^d}} \tag{1.66}$$

or, in components,

$$u^\mu = \frac{g^{0\mu}}{\sqrt{|g_{00}|}}. \tag{1.67}$$

This equation becomes invalid, and static observers no longer exist, when r approaches the static limit r_{SL}. When $r \leq r_{\text{SL}}$ all observers are forced to co-rotate with the spacetime, a phenomenon known as the "dragging of inertial frames".

The static limit is not an event horizon; there is an event horizon located at

$$r_+ = m + \sqrt{m^2 - a^2}, \tag{1.68}$$

where the quantity Δ vanishes. The static limit touches the event horizon at the poles $\theta = \pm\pi/2$, where $r_{\text{SL}} = r_+$, and the region between the static limit and the event horizon is called *ergosphere*. The event horizon exists only if $a \leq m$, equivalent to $J \leq m^2$. If $a = m$ (or $J = m^2$), the black hole is *extremal*. For $a > m$, the Kerr metric describes a naked singularity and the extremal black hole, therefore, constitutes a threshold between black holes and naked singularities in parameter space.

There are two roots of $g^{11} = 0$, the equation which, as we will see later, locates the horizons. Using Eq. (1.61), this condition is seen to be equivalent to $\Delta = r^2 - 2mr + a^2 = 0$, which has as roots the radius of the event horizon $r = r_+$ and

$$r_- = m - \sqrt{m^2 - a^2}. \tag{1.69}$$

This surface is analogous to the inner horizon of the Reissner-Nordström spacetime. The two horizons at $r = r_\pm$ coincide for an extremal black hole with $a = m$.

1.6   Kerr-Newman Metric

The Kerr-Newman spacetime is interpreted as describing an electrically charged spinning black hole. It is a three-parameter (mass M, angular momentum per unit mass a, and electric charge Q) family of metrics. The *Kerr-Newman line element in Boyer-Linquist coordinates* is

$$ds^2 = -\frac{\rho^2 \Delta}{\Sigma} dt^2 + \frac{\Sigma}{\rho^2} \sin^2 \theta \, (d\varphi - \omega dt)^2 + \frac{\rho^2}{\Delta} dr^2 + \rho^2 d\theta^2 \qquad (1.70)$$

where

$$\rho^2 = r^2 + a^2 \cos^2 \theta \,, \qquad (1.71)$$

$$\Delta = r^2 - 2mr + a^2 + Q^2 \,, \qquad (1.72)$$

$$\Sigma = \left(r^2 + a^2\right)^2 - a^2 \Delta \sin^2 \theta \,, \qquad (1.73)$$

$$\omega = \frac{a\left(r^2 + a^2 - \Delta\right)}{\Sigma} \,. \qquad (1.74)$$

The Kerr-Newman metric reduces to the Kerr metric in the limit $Q \to 0$ and to the Reissner-Nordström one when $a \to 0$. The static limit is given by

$$r_{\mathrm{SL}}(\theta) = m + \sqrt{m^2 - Q^2 - a^2 \cos^2 \theta} \,; \qquad (1.75)$$

the horizon is located at

$$r_+ = m + \sqrt{m^2 - Q^2 - a^2} \,, \qquad (1.76)$$

with the ergosphere comprising the region $r_+ < r < r_{\mathrm{SL}}$.

1.7   Energy Conditions

Here we summarize the point-wise energy conditions of General Relativity. The energy conditions satisfied by a certain form of matter are formulated in terms of its energy-momentum tensor T_{ab}. In order to visualize an energy condition, it is useful to see the form that it assumes for a perfect fluid characterized by the stress-energy tensor

$$T_{ab} = (P + \rho) \, u_a u_b + P g_{ab} \,. \qquad (1.77)$$

- The *weak energy condition* (WEC) is

$$T_{ab}\, t^a\, t^b \geq 0 \qquad \text{for all timelike vectors } t^a. \qquad (1.78)$$

For the fluid (1.77), this condition becomes

$$\rho \geq 0 \quad \text{and} \quad \rho + P \geq 0. \qquad (1.79)$$

- The *dominant energy condition* (DEC) consists of the WEC and of the extra requirement that $T^{ab}\, t_a$ be a null or timelike vector (i.e., $T_{ab}T^b{}_c\, t^a\, t^c \leq 0$) for any timelike vector t^a. For the fluid (1.77) the DEC assumes the form

$$\rho \geq |P|, \qquad (1.80)$$

(i.e., the speed of the energy flow does not exceed the speed of light).
- The *null energy condition* (NEC) consists of

$$T_{ab}\, l^a\, l^b \geq 0 \qquad \text{for all null vectors } l^a; \qquad (1.81)$$

for the fluid (1.77), this means that

$$\rho + P \geq 0. \qquad (1.82)$$

Violations of the NEC are studied in the context of macroscopic traversable wormholes and occur in cosmology if the expansion of the universe is super-accelerated ($\dot{H} > 0$, where H is the Hubble parameter); in this case energy is called *phantom energy*.
- The *null dominant energy condition* (NDEC) consists of

$$T_{ab}\, l^a\, l^b \geq 0 \quad \text{and} \quad T^{ab}\, l_b \text{ is null or timelike for any null vector } l^a. \qquad (1.83)$$

The NDEC resembles the DEC but here l^a is null instead of timelike. The NDEC for the fluid (1.77) amounts to

$$\rho \geq |P| \quad \text{or} \quad \rho = -P. \qquad (1.84)$$

- The *strong energy condition* (SEC) consists of

$$\left(T_{ab} - \frac{1}{2}Tg_{ab}\right) t^a\, t^b \geq 0 \quad \text{for any timelike vector } t^a \qquad (1.85)$$

or, for the fluid (1.77),

$$\rho + P \geq 0 \quad \text{and} \quad \rho + 3P \geq 0. \qquad (1.86)$$

This condition ensures that gravity is always attractive and is violated by a positive cosmological constant (which satisfies $P^{(\Lambda)} = -\rho^{(\Lambda)}$), during inflation, or in a dark energy-dominated cosmological era with $P < -\rho/3$.

Quantum systems are expected to violate all of the energy conditions, including positivity of the energy density, on short timescales (e.g., [22]). However, while a negative energy density is permitted for a quantum system during a certain interval of time, it appears that later on the system more than compensates with a positive energy density (*quantum interest*), thus respecting some averaged energy conditions.

Even at the classical level, alternative theories of gravity and the theory of a scalar field nonminimally coupled to the Ricci curvature contain fields which can violate all of the energy conditions [10, 20].

1.8 Conclusions

The previous list of spacetimes commonly associated with black holes is not exhaustive. For detailed discussions of these spacetimes, the timelike and null geodesics in them, and the thermodynamics of their horizons, see Refs. [11, 23, 48, 58, 71, 74, 75]. In the following chapters we assume that the reader has some familiarity with these basic solutions of Einstein's theory and we will focus on causal barriers which preclude the knowledge of spacetime regions to families of (timelike) observers.

Problems

1.1. Prove that $u \equiv t - r^*$ and $v \equiv t + r^*$ are null coordinates. Show that the coordinates $\bar{u} \equiv t - r$ and $\bar{v} \equiv t + r$, instead, are not null.

1.2. Establish the causal character of the inner and outer horizons of the Reissner-Nordström metric (1.44).

References

1. Amendola, L., Tsujikawa, S.: Dark Energy, Theory and Observations. Cambridge University Press, Cambridge (2010)
2. Arnowitt, R.L., Deser, S., Misner, C.W.: The dynamics of general relativity. In: Witten, L. (ed.) Gravitation: An Introduction to Current Research, pp. 227–265. Wiley, New York (1962). [Reprinted in arXiv:gr-qc/0405109]
3. Ashtekar, A., Galloway, G.J.: Some uniqueness results for dynamical horizons. Adv. Theor. Math. Phys. **9**, 1 (2005)

4. Ashtekar, A., Krishnan, B.: Isolated and dynamical horizons and their applications. Living Rev. Relat. **7**, 10 (2004)
5. Babichev, E., Charmousis, C.: Dressing a black hole with a time-dependent Galileon. J. High Energy Phys. **1408**, 106 (2014)
6. Baumgarte, T.W., Shapiro, S.L.: Numerical relativity and compact binaries. Phys. Rept. **376**, 41 (2003)
7. Booth, I.: Black hole boundaries. Can. J. Phys. **83**, 1073 (2005)
8. Buchdahl, H.A.: Isotropic coordinates and Schwarzschild metric. Int. J. Theor. Phys. **24**, 731 (1985)
9. Burko, L.M., Ori, A. (eds.): Internal Structure of Black Holes and Spacetime Singularities, An International Research Workshop, Haifa (IOP, Bristol, 1997)
10. Capozziello, S., Faraoni, V.: Beyond Einstein Gravity, A Survey of Gravitational Theories for Cosmology and Astrophysics (Springer, New York, 2010)
11. Carroll, S.M.: Spacetime and Geometry—An Introduction to General Relativity (Addison-Wesley, San Francisco, 2004)
12. Carter, B.: Axisymmetric black hole has only two degrees of freedom. Phys. Rev. Lett. **26**, 331 (1970)
13. Chu, T., Pfeiffer, H.P., Cohen, M.I.: Horizon dynamics of distorted rotating black holes. Phys. Rev. D **83**, 104018 (2011)
14. Collins, W.: Mechanics of apparent horizons. Phys. Rev. D **45**, 495 (1992)
15. Davis, A.-C., Gregory, R., Jha, R., Muir, J.: Astrophysical black holes in screened modified gravity. J. Cosmol. Astropart. Phys. **1408**, 033 (2014)
16. Detweiler, S.: Resource letter BH-1: black holes. Am. J. Phys. **49**, 394 (1981)
17. De Felice, A., Tsujikawa, S.: $f(R)$ theories. Living Rev. Relat. **13**, 3 (2010)
18. Eddington, A.S.: A comparison of Whitehead's and Einstein's formulas. Nature **113**, 192 (1924)
19. Eling, C., Guedens, R., Jacobson, T.: Nonequilibrium thermodynamics of spacetime. Phys. Rev. Lett. **96**, 121301 (2006)
20. Faraoni, V.: Cosmology in Scalar-Tensor Gravity. Kluwer Academic, Dordrecht (2004)
21. Finkelstein, D.: Past-future asymmetry of the gravitational field of a point particle. Phys. Rev. D **110**, 965 (1958)
22. Ford, L.H., Roman, T.A.: Classical scalar fields and violations of the second law. Phys. Rev. D **64**, 024023 (2001)
23. Frolov, V.P., Novikov, I.D.: Black Hole Physics, Basic Concepts and New Developments. Kluwer Academic, Dordrecht (1998)
24. Gallo, E., Marolf, D.: Resource letter BH-2: black holes. Am. J. Phys. **77**, 294 (2009)
25. Gautreau, R.: Light cones inside the Schwarzschild radius. Am. J. Phys. **63**, 431 (1995)
26. Gautreau, R., Hoffmann, B.: The Schwarzschild radial coordinate as a measure of proper distance. Phys. Rev. D **17**, 2552 (1978)
27. Geroch, R.: Limits of spacetimes. Commun. Math. Phys. **13**, 180 (1969)
28. Gibbons, G.W., Hawking, S.W.: Cosmological event horizon, thermodynamics, and particle creation. Phys. Rev. D **15**, 2738 (1977)
29. Gullstrand, A.: Allgemeine lösung de statischen eink örper-problems in der Einsteinschen gravitations theories. Ark. Mat. Astron. Fys. **16**, 1 (1922)
30. Gundlach, C., Price, R.H., Pullin, J.: Late-time behavior of stellar collapse and explosions. I. Linearized perturbations. Phys. Rev. D **49**, 883 (1994)
31. Gürses, M., Gürsey, F.: Lorentz covariant treatment of the Kerr-Schild geometry. J. Math. Phys. **16**, 2385 (1975)
32. Hawking, S.W.: Black hole explosions? Nature **248**, 30 (1970)
33. Hawking, S.W.: Black holes in general relativity. Commun. Math. Phys. **25**, 152 (1972)
34. Hawking, S.W.: Particle creation by black holes. Commun. Math. Phys. **43**, 199 (1975); Erratum **46**, 206 (1976)
35. Hayward, S.A.: General laws of black hole dynamics. Phys. Rev. D **49**, 6467 (1994)

36. Herdeiro, C.A.R., Radu, E.: Kerr black holes with scalar hair. Phys. Rev. Lett. **112**, 221101 (2014)
37. Israel, W.: Event horizons in static vacuum space-times. Phys. Rev. **164**, 1776 (1967)
38. Jacobson, T.: Thermodynamics of spacetime: the Einstein equation of state. Phys. Rev. Lett. **75**, 1260 (1995)
39. Kerr, R.P.: Gravitational field of a spinning mass as an example of algebraically special metrics. Phys. Rev. Lett. **11**, 237 (1963)
40. Kerr, R.P., Schild, A.: A new class of vacuum solutions of the Einstein field equations. In: Atti del Convegno sulla relatività generale: problemi dell'energia e onde gravitazionali, p. 222. Barbera, Firenze (1965)
41. Kerr, R.P., Schild, A.: Some algebraically degenerate solutions of Einstein's gravitational field equations. Proc. Symp. Appl. Math. **17**, 199 (1965)
42. Kolb, E.W., Turner, M.S.: The Early Universe. Addison-Wesley, Reading (1990)
43. Kraus, P., Wilczek, F.: Some applications of a simple stationary line element for the Schwarzschild geometry. Mod. Phys. Lett. A **9**, 3713 (1995)
44. Kruskal, M.D.: Maximal extension of Schwarzschild metric. Phys. Rev. **119**, 1743 (1960)
45. Lake, K.: A class of quasi-stationary regular line elements for the Schwarzschild geometry. Preprint, arXiv:gr-qc/9407005
46. Liddle, A.R., Lyth, D.H.: Cosmological Inflation and Large Scale Structure. Cambridge University Press, Cambridge (2000)
47. Martel, K., Poisson, E.: Regular coordinate systems for Schwarzschild and other spherical spacetimes. Am. J. Phys. **69**, 476 (2001)
48. Misner, C.W., Thorne, K.S., Wheeler, J.A.: Gravitation. Freeman, New York (1973)
49. Mukhanov, V.: Physical Foundations of Cosmology. Cambridge University Press, Cambridge (2005)
50. Muller zum Hagen, H., Robinson, D.C., Seifert, H.J.: Black holes in static vacuum space-times. Gen. Rel. Gravit. **4**, 53 (1973)
51. Nielsen, A.B.: Black holes and black hole thermodynamics without event horizons. Gen. Rel. Gravit. **41**, 1539 (2009)
52. Nielsen, A.B., Visser, M.: Production and decay of evolving horizons. Class. Quantum Grav. **23**, 4637 (2006)
53. Novikov, I.D.: PhD thesis, Shternberg Astronomical Institute, Moscow (1963)
54. Painlevé, P.: La méchanique classique et la théorie de la relativité. Comp. Rend. Acad. Sci. (Paris) **173**, 677 (1921)
55. Paiva, F.M., Romero, C.: On the limits of Brans-Dicke space-times: a coordinate-free approach. Gen. Rel. Gravit. **25**, 1305 (1993)
56. Paiva, F.M., Reboucas, M., MacCallum, M.: On limits of space-times: a coordinate-free approach. Class. Quantum Grav. **10**, 1165 (1993)
57. Paiva, F.M., Reboucas, M., Hall, G.S., MacCallum, M.: Limits of the energy-momentum tensor in general relativity. Class. Quantum Grav. **15**, 1031 (1998)
58. Poisson, E.: A Relativist's Toolkit: The Mathematics of Black-Hole Mechanics. Cambridge University Press, Cambridge (2004)
59. Poisson, E., Israel, W.: The internal structure of black holes. Phys. Rev. D **41**, 1796 (1990)
60. Price, R.H.: Nonspherical perturbations of relativistic gravitational collapse. I. Scalar and gravitational perturbations. Phys. Rev. D **5**, 2419 (1972)
61. Price, R.H.: Nonspherical perturbations of relativistic gravitational collapse. II. Integer-spin, zero-rest-mass fields. Phys. Rev. D **5**, 2439 (1972)
62. Regge, T., Wheeler, J.A.: Stability of a Schwarzschild singularity. Phys. Rev. **108**, 1063 (1957)
63. Rinaldi, M.: Black holes with nonminimal derivative coupling. Phys. Rev. D **86**, 084048 (2012)
64. Rindler, W.: Visual horizons in world-models. Mon. Not. R. Astr. Soc. **116**, 663 (1956). [Reprinted in Gen. Rel. Gravit. **34**, 133 (2002)]
65. Robertson, H.P., Noonan, T.W.: Relativity and Cosmology. Saunders, Philadelphia (1968)
66. Robinson, D.C.: Classification of black holes with electromagnetic fields. Phys. Rev. D **10**, 458 (1974)

67. Robinson, D.C.: Uniqueness of the Kerr black hole. Phys. Rev. Lett. **34**, 905 (1975)
68. Sotiriou, T.P., Faraoni, V.: $f(R)$ theories of gravity. Rev. Mod. Phys. **82**, 451 (2010)
69. Sotiriou, T.P., Zhou, S.-Y.: Black hole hair in generalized scalar-tensor gravity. Phys. Rev. Lett. **112**, 251102 (2014)
70. Sotiriou, T.P., Zhou, S.-Y.: Black hole hair in generalized scalar-tensor gravity: an explicit example. Phys. Rev. D **90**, 124063 (2014)
71. Stephani, H., Kramer, D., MacCallum, M., Hoenselaers, C., Herlt, E.: Exact Solutions of Einstein's Field Equations, 2nd edn. Cambridge University Press, Cambridge (2003)
72. Szekeres, G.: On the singularities of a Riemannian manifold. Publ. Mat. Debr. **7**, 285 (1960)
73. Thornburg, J.: Event and apparent horizon finders for $3 + 1$ numerical relativity. Living Rev. Relat. **10**, 3 (2007)
74. Wald, R.M.: General Relativity. Chicago University Press, Chicago (1984)
75. Wald, R.M.: The thermodynamics of black holes. Living Rev. Relat. **4**, 6 (2001)
76. Weyl, H.: Zur Gravitationstheorie. Ann. Phys. (Leipzig) **54**, 117 (1917)
77. Wheeler, J.A.: Geons. Phys. Rev. **97**, 511 (1955)

Chapter 2
Horizons

The greatest deception men suffer is from their own opinions.

—Leonardo da Vinci

2.1 Introduction

We will now review standard tools used in the analysis of horizons and black hole physics, most notably the congruences of null geodesics crossing a horizon. Many textbooks present this standard material but we will recall it here anyway to provide a more self-contained discussion. Then we will present the laws describing how these null geodesic congruences are affected by gravity. Certain formulae which are used in the calculations of the following chapters will also be introduced here. The case of spherical symmetry is particularly important because most of the known analytic dynamical solutions of Einstein theory and of alternative theories of gravity, which are used to gain physical insight into both gravitational physics and the physics of horizons, are spherically symmetric.

2.2 Null Geodesic Congruences and Trapped Surfaces

A *null geodesic* is a curve on the spacetime manifold which has null tangent l^a (i.e., $l^a l_a = 0$) and satisfies the *geodesic equation*

$$l^b \nabla_b l^a = \alpha(\lambda) \, l^a \,, \tag{2.1}$$

where λ is a parameter along the curve. The geodesic equation expresses the fact that the tangent is transported parallel to itself as one moves along the geodesic, or, the fact that this curve is "as straight as possible" in the curved spacetime.

The parameter λ can be chosen so that the geodesic equation simplifies to

$$l^b \nabla_b l^a = 0 \tag{2.2}$$

© Springer International Publishing Switzerland 2015
V. Faraoni, *Cosmological and Black Hole Apparent Horizons*,
Lecture Notes in Physics 907, DOI 10.1007/978-3-319-19240-6_2

or, in components,

$$\frac{d^2 x^\mu}{d\lambda^2} + \Gamma^\mu_{\alpha\beta} \frac{dx^\alpha}{d\lambda} \frac{dx^\beta}{d\lambda} = 0, \tag{2.3}$$

where $x^\mu(\lambda)$ are the coordinates of points along the curve. If the geodesic equation reduces to the simple form (2.3), we say that it is *affinely parametrized* and that λ is an *affine parameter*. Two affine parameters λ and λ' can differ at most by an affine transformation,

$$\lambda' = a\lambda + b \tag{2.4}$$

where a and b are real constants (e.g., [78]).

Let O be an open region of the spacetime manifold; a *congruence of curves* in O is a family of curves such that through every point of O passes one and only one curve of the family. The tangents to these curves define a vector field on O (and, conversely, every continuous vector field in O generates a congruence of curves, those to which the vector field is tangent). If the field of tangents is smooth, we say that the congruence is smooth. In particular, we can consider a *congruence of null geodesics* with tangents l^a in the open region O.

Consider a congruence of affinely parametrized null geodesics with tangent l^a and affine parameter λ, which satisfy $l^a l_a = 0$ and $l^b \nabla_b l^a = 0$. Let us consider now another parameter s which labels the various geodesics of the congruence in O, so that the family of curves parameterized by λ and s generates a 2-dimensional surface with coordinates λ and s. Points along the geodesics of the congruence in O have spacetime coordinates $x^\mu = x^\mu(\lambda, s)$ and, in analogy with the tangent $l^\mu = \partial x^\mu / \partial \lambda$, we can introduce the *deviation vector* with components $\eta^\mu \equiv \partial x^\mu / \partial s$. By construction, it is

$$\eta^a l_a = 0 \tag{2.5}$$

but η^a can still have a component along the curves, that is, parallel to l^a because l^a is null (this would not be the case if we were considering timelike geodesics instead). However, we can restrict ourselves to deviation vectors which are considered to be equivalent if they differ only by a component along l^a. More precisely: define the equivalence relation $\eta^a \sim \eta^{a'} \Leftrightarrow \eta^{a'} - \eta^a = b l^a$ for some real number b. It is straightforward to show that $\sim$ is an equivalence relation and we can consider equivalence classes of deviation vectors.

The tangent space composed of all vectors orthogonal in this sense to l^a constitutes a 2-dimensional vector space and we can consider its dual and the space of tensors built with them (see Refs. [78, 95] for details). It can be proved that the geodesic deviation vector η^a satisfies the *geodesic deviation equation*

$$\frac{D^2 \eta^a}{D\lambda^2} = -R^a_{\ bcd} u^b \eta^c u^d, \tag{2.6}$$

which expresses the fact that neighbouring geodesics deviate from each other because of the spacetime curvature.

Consider now a congruence of null geodesics with tangent l^a parametrized by an affine parameter λ, with $l_a l^a = 0$ and $l^c \nabla_c l^a = 0$. The metric h_{ab} in the 2-space orthogonal to l^a is constructed as follows [78]: select another null vector field n^a such that $n_c n^c = 0$. The null vectors l^c and n^c are defined up to rescalings; here we normalize them according to $l^c n_c = -1$, but other choices are possible and will be used later. Then, the 2-metric orthogonal to l^a is

$$h_{ab} \equiv g_{ab} + l_a n_b + l_b n_a . \tag{2.7}$$

It is easy to verify that h_{ab} is purely spatial and $h^a{}_b$ is a projection operator onto the 2-space orthogonal to l^a:

$$h_{ab} l^a = h_{ab} l^b = 0 , \tag{2.8}$$

$$h^a{}_a = 2 , \tag{2.9}$$

$$h^a{}_c h^c{}_b = h^a{}_b . \tag{2.10}$$

Only the null congruence with tangent l^a is fixed and the choice of n^a is not unique, but geometric and physically relevant quantities do not depend on it.

Let η^a be the geodesic deviation vector and define the tensor field [78, 95]

$$B_{ab} \equiv \nabla_b l_a , \tag{2.11}$$

which satisfies $l^b \nabla_b \eta^a = B^a{}_b \eta^b$ and is orthogonal to the null geodesics, $B_{ab} l^a = B_{ab} l^b = 0$. The transverse part of the deviation vector ("relative velocity" of two neighbouring geodesics) is

$$\tilde{\eta}^a \equiv h^a{}_b \eta^b = \eta^a + (n^c \eta_c) l^a . \tag{2.12}$$

The orthogonal component of $l^c \nabla_c \eta^a$, denoted by a tilde, is [78]

$$\widetilde{(l^c \nabla_c \eta^a)} = h^a{}_b h^c{}_d B^b{}_c \tilde{\eta}^d \equiv \tilde{B}^a{}_d \tilde{\eta}^d . \tag{2.13}$$

The transverse tensor $\tilde{B}_{ab}$ is decomposed into its symmetric and antisymmetric parts, and the symmetric part is further decomposed into its trace and trace-free parts as [78, 95]

$$\tilde{B}_{ab} = \tilde{B}_{(ab)} + \tilde{B}_{[ab]} \equiv \left(\frac{\theta}{2} h_{ab} + \sigma_{ab} \right) + \omega_{ab} , \tag{2.14}$$

where the trace

$$\theta \equiv g^{ab} \tilde{B}_{ab} = g^{ab} B_{ab} = \nabla_c l^c \tag{2.15}$$

is the *expansion* of the affinely-parametrized congruence (this quantity does not depend on the vector n^a);

$$\theta_{ab} \equiv \frac{\theta}{2} h_{ab} \qquad (2.16)$$

is the *expansion tensor*;

$$\sigma_{ab} \equiv \tilde{B}_{(ab)} - \frac{\theta}{2} h_{ab} \qquad (2.17)$$

is the *shear tensor*; and

$$\omega_{ab} \equiv \tilde{B}_{[ab]} \qquad (2.18)$$

is the *vorticity tensor*. The expansion, shear, and vorticity tensors are purely transversal:

$$\theta_{ab} l^a = \theta_{ab} l^b = 0 , \qquad (2.19)$$

$$\sigma_{ab} l^a = \sigma_{ab} l^b = 0 , \qquad (2.20)$$

$$\omega_{ab} l^a = \omega_{ab} l^b = 0 , \qquad (2.21)$$

and the shear and vorticity are trace-free, $\sigma^a{}_a = \omega^a{}_a = 0$. The *shear scalar* and *vorticity scalar* are

$$\sigma^2 \equiv \sigma_{ab} \sigma^{ab} , \qquad \omega^2 \equiv \omega_{ab} \omega^{ab} \qquad (2.22)$$

and they are non-negative. The propagation of the expansion along a null geodesic is ruled by the *Raychaudhuri equation* [78, 95]

$$\frac{d\theta}{d\lambda} = -\frac{\theta^2}{2} - \sigma^2 + \omega^2 - R_{ab} l^a l^b \qquad (2.23)$$

(which is invariant under redefinitions of n^a), and similar propagation equations hold for σ_{ab} and ω_{ab} [95]. This equation describes how null rays are focused ($d\theta/d\lambda < 0$) or defocused ($d\theta/d\lambda > 0$) by expansion itself, shear, rotation, and matter (related to R_{ab} through the Einstein equations). Conventional forms of energy in General Relativity satisfy the positive curvature condition $R_{ab} l^a l^b \geq 0$ and the old adage "gravity always focuses" follows from this condition. Therefore, a gravitational lens will focus light rays passing nearby. Shear acts as gravity while rotation acts in the opposite direction. This effect is familiar in Newtonian gravity, where the rotation of a massive body is associated with a centrifugal acceleration which counteracts the gravitational attraction (except at the poles).

Thus far, we have considered affinely-parametrized geodesics. If the congruence of null geodesics with tangent l^a is not affinely-parametrized, the geodesic equation assumes the form

$$l^c \nabla_c l^a = \kappa \, l^a \qquad\qquad (2.24)$$

instead of $l^c \nabla_c l^a = 0$, where the quantity κ which measures the failure of l^a to be affinely-parametrized is sometimes used, on a horizon, as a possible definition of *surface gravity* [72, 77]. The expansion of a congruence of non-affinely parametrized null geodesics is then

$$\theta = \nabla_c l^c - \kappa \qquad\qquad (2.25)$$

and the Raychaudhuri equation becomes [78]

$$\frac{d\theta}{d\lambda} = \kappa\,\theta - \frac{\theta^2}{2} - \sigma^2 + \omega^2 - R_{ab} l^a l^b . \qquad (2.26)$$

A compact and orientable surface has two independent directions orthogonal to it, corresponding to ingoing and outgoing null rays. In the presence of spherical symmetry, one is naturally led to study congruences of *radial* ingoing and outgoing null geodesics with tangent fields l^a and n^a, respectively, which are orthogonal to the 2-spheres of symmetry. In this case the role of the vector n^a is played by the tangent to the ingoing null geodesics (hence the use of the same symbol). To compute the expansion of the null vector l^a when the geodesic to which it is tangent is not necessarily affinely-parametrized, the relation

$$\theta_l = h^{ab} \nabla_a l_b = \left[g^{ab} + \frac{l^a n^b + n^a l^b}{(-n^c l^d g_{cd})} \right] \nabla_a l_b \qquad (2.27)$$

is useful. Here $h_a{}^b$ acts as a projection tensor onto the two-dimensional surface to which both l^a and n^a are normal. If l^a is null everywhere then the expression $l^b \nabla_a l_b$ in the third term on the right hand side vanishes identically. Equation (2.27) is independent of the field equations of the theory of gravity and can be applied when l^c and n^c are not normalized to satisfy $l^c n_c = -1$ (this normalization is usually imposed but it is not necessary and it sometimes conflicts with other requirements that one may want to impose on l^c and n^c, such as in the various possible definitions of surface gravity on a horizon [72] or in calculations of quasi-local energy [49]).

The following are basic definitions for closed 2-surfaces in regard to the ingoing and outgoing null geodesic congruences,[1] which have expansions θ_n and θ_l, respectively [4, 21, 24, 68]:

- A *normal surface* corresponds to $\theta_l > 0$ and $\theta_n < 0$ (this is the case, e.g., of a 2-sphere in Minkowski space in the absence of gravity).

---

[1]While these 2-surfaces are usually assumed to be spacelike [4, 19, 21] this condition is not imposed here.

- A *trapped surface* [76] corresponds to $\theta_l < 0$ and $\theta_n < 0$. In this case, the *outgoing*, in addition to the usual ingoing, future-directed null rays are converging instead of diverging—light propagating outward is dragged back by strong gravity.
- A *marginally outer trapped* (or *marginal*) *surface (MOTS)* corresponds to $\theta_l = 0$ (where l^a is the outgoing null normal to the surface) and $\theta_n < 0$.
- An *untrapped surface* is one with $\theta_l \theta_n < 0$.
- An *anti-trapped surface* corresponds to $\theta_l > 0$ and $\theta_n > 0$ (both outgoing and ingoing future-directed null rays are diverging).
- A *marginally outer trapped tube (MOTT)* is a 3-dimensional surface which can be foliated entirely by marginally outer trapped (2-dimensional) surfaces.

It was proved by Penrose that, in General Relativity, if a spacetime contains a trapped surface, the null energy condition holds, and there is a non-compact Cauchy surface for the spacetime, then the spacetime contains a singularity [76]. Trapped surfaces seem to be essential features in the black hole concept and notions of "horizon" of practical utility will be identified with the boundaries of spacetime regions which contain trapped surfaces. The mathematical conditions for the existence and uniqueness of MOTSs are not totally clear at the moment. In general, a marginally outer trapped tube can be distorted smoothly, which implies that MOTTs are non-unique [2, 21, 39].

We will now examine various types of horizons which appear in the literature on black holes, cosmology, quantum field theory in curved spaces, and the thermodynamics usually associated with these horizons.

2.3 Rindler Horizons for Accelerated Observers in Minkowski Spacetime

In Minkowski space, consider a particle of position $\mathbf{x}(t)$ moving with 3-velocity $\mathbf{u} = d\mathbf{x}/dt$ in an inertial frame. Its 3-dimensional acceleration is $\mathbf{a} \equiv d\mathbf{u}/dt$. The transformation law of the 3-acceleration between this frame and another inertial frame moving with constant speed v with respect to it is [87]

$$a'_x = \frac{1}{\gamma^3} \frac{a_x}{(1 - vu_x)^3}, \tag{2.28}$$

$$a'_y = \frac{1}{\gamma^2 (1 - vu_x)^2} \left[a_y + \frac{a_x vu_y}{(1 - vu_x)} \right], \tag{2.29}$$

$$a'_z = \frac{1}{\gamma^2 (1 - vu_x)^2} \left[a_z + \frac{a_x vu_z}{(1 - vu_x)} \right], \tag{2.30}$$

where $\gamma = \left(1 - v^2\right)^{-1/2}$ is the Lorentz factor. Let us define now *uniform acceleration*: for simplicity, assume that the particle moves along the x-axis, so $\mathbf{u} = (u, 0, 0)$.

Clearly, one cannot proceed as in Newtonian theory by requiring that $du/dt = $ constant $\equiv a$ for constant acceleration because then $u(t) = u_0 + at$ exceeds the speed of light when t is sufficiently large, which is not possible in Special Relativity. Therefore, a different definition is required (e.g., Ref. [38]). *A particle has uniform acceleration if and only if its acceleration has the same value at each instant in any inertial frame comoving with the particle* (i.e., in any frame moving with the same velocity as the particle, in which the particle is instantaneously at rest). At different times there are different inertial frames which are comoving but it is required that they all measure the same acceleration a of the particle. In other words, it is required that the particle move along a straight line and has *constant proper acceleration a* (where the proper acceleration is the particle acceleration in the frame in which the particle is instantaneously at rest).

In any instantaneously comoving frame (characterized by $v = u$), the particle has velocity and acceleration

$$u' = 0 , \tag{2.31}$$

$$\frac{du'}{dt'} = a = \text{const.} \tag{2.32}$$

Then, the transformation rule for the acceleration between inertial frames gives (we invert Eq. (2.28) by exchanging primed and unprimed quantities and changing v into $-v$)

$$\frac{du}{dt} = \frac{1}{\gamma^3 (1 + vu')^3} \frac{du'}{dt'} \tag{2.33}$$

and, setting $u' = 0$ and $du'/dt' = a = $ constant,

$$\left(1 - u^2\right)^{-3/2} \frac{du}{dt} = a . \tag{2.34}$$

This equation is integrated by remembering that $\displaystyle\int \frac{dz}{\sqrt{\left(\alpha^2 - z^2\right)^3}} = \frac{z}{\alpha^2 \sqrt{\alpha^2 - z^2}}$, yielding the algebraic equation for u

$$\frac{u}{\sqrt{1 - u^2}} = a (t - t_0) \tag{2.35}$$

where we assumed that an integration constant u_0 vanishes. This algebraic equation is solved as

$$\frac{dx}{dt} = u = \frac{a (t - t_0)}{\sqrt{1 + a^2 (t - t_0)^2}} , \tag{2.36}$$

and this ordinary differential equation can be easily integrated to yield

$$x(t) - x_0 = \frac{1}{a}\sqrt{1 + a^2 (t - t_0)^2} - \frac{1}{a}. \tag{2.37}$$

By squaring and collecting terms, one obtains

$$\frac{\left(x - x_0 + \frac{1}{a}\right)^2}{\left(\frac{1}{a}\right)^2} - \frac{(t - t_0)^2}{\left(\frac{1}{a}\right)^2} = 1. \tag{2.38}$$

For simplicity we take the initial condition $x_0 = 1/a$ at $t_0 = 0$, reducing Eq. (2.38) to

$$x^2 - t^2 = \left(\frac{1}{a}\right)^2, \tag{2.39}$$

a family of hyperbolae parametrized by the constant acceleration a and called *hyperbolic motions* [25] or worldlines of *Rindler observers* [38]. The worldline of the uniformly accelerated observer can be parametrized by its proper time τ and has equation

$$x^\mu(\tau) = \left(t(\tau), x(\tau), 0, 0\right) = \left(\frac{\sinh(a\tau)}{a}, \frac{\cosh(a\tau)}{a}, 0, 0\right) \tag{2.40}$$

and tangent

$$u^\mu = \left(\cosh(a\tau), \sinh(a\tau), 0, 0\right) \tag{2.41}$$

with $u_c u^c = -1$. Then, it is

$$v = u = \frac{dx}{dt} = \tanh(a\tau) \tag{2.42}$$

and the Lorentz factor is

$$\gamma = \cosh(a\tau). \tag{2.43}$$

As is evident by writing $t = \pm\sqrt{x^2 - \frac{1}{a^2}}$, these hyperbolae have the branches of the light cone through the origin $t = \pm x$ as asymptotes. The effect of choosing arbitrary initial condition x_0 and arbitrary initial time t_0 is only to move the origin. Each hyperbola is the worldline of a uniformly accelerated observer travelling in the x-direction. It is clear from Fig. 2.1 that the lines $t = \pm x$ separate the spacetime in two regions: the region to the left of this null cone through the origin is forever unaccessible to the uniformly accelerated observer because signals sent from it would have to travel faster than light to cross the line $t = x$ and reach the observer. This observer has an horizon, called *Rindler* (or *acceleration*) *horizon*. An horizon is

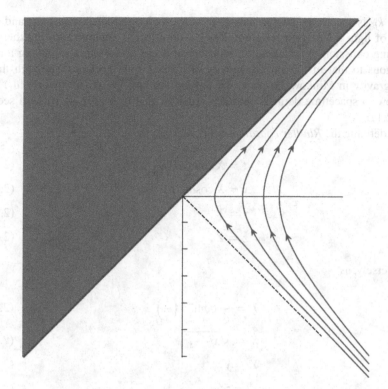

Fig. 2.1 The Rindler horizon of a uniformly accelerated observer is the boundary of the shaded region, i.e., the line $t = x$. A light signal emitted from the shaded region would have to travel faster than light to reach the accelerated observer to the right of it

a causal barrier, a surface which separates spacetime into two regions—the region beyond the horizon cannot causally influence the region in which the observer is located and no information sent by the region beyond the horizon can ever leak out and reach the observer. Vice-versa, signals sent by the observer can cross the event horizon and propagate through the other region. The location of the event horizon depends on the uniformly accelerated observer: different accelerated observers will determine different acceleration horizons.

A uniformly accelerated observer in Minkowski spacetime is subject to the *Unruh effect* of quantum field theory [34, 44, 89] (see, e.g., [27] for a pedagogical exposition). Although an observer at rest in this Minkowski spacetime "sees" a quantum vacuum corresponding to zero particles of a quantum field (usually taken, for simplicity, to be a massless minimally coupled scalar field), the uniformly accelerated observer will detect a thermal bath of particles in equilibrium at the *Unruh temperature* given by

$$k_B T = \left(\frac{\hbar}{c}\right) \frac{a}{2\pi}, \qquad (2.44)$$

where k_B is the Boltzmann constant, $\hbar$ is the reduced Planck constant, and the speed of light c has been restored. The Unruh effect is analogous to the thermal Hawking emission by black holes, with the uniform acceleration a playing a role analogous to that of the surface gravity at a black hole horizon. Although there is no gravity in Minkowski space, the Killing vector $\xi^a = x(\partial/\partial t)^a + t(\partial/\partial_x)^a$ of Minkowski spacetime defines a surface gravity equal to a [27], as we will see in Sect. 2.12.

By defining the *Rindler coordinates* (T,X,Y,Z) as

$$t = X\sinh(aT)\,, \tag{2.45}$$

$$x = X\cosh(aT)\,, \tag{2.46}$$

$$y = Y\,, \tag{2.47}$$

$$z = Z\,, \tag{2.48}$$

or, inversely, as

$$T = \frac{1}{a}\tanh^{-1}\left(\frac{t}{x}\right)\,, \tag{2.49}$$

$$X = \sqrt{x^2 - t^2}\,, \tag{2.50}$$

$$Y = y\,, \tag{2.51}$$

$$Z = z\,, \tag{2.52}$$

for $|t| < |x|$, the Minkowski line element $ds^2 = -dt^2 + dx^2 + dy^2 + dz^2$ is turned into the *Rindler line element*

$$ds^2 = -a^2 X^2 dT^2 + dX^2 + dY^2 + dZ^2\,. \tag{2.53}$$

The Rindler coordinate chart covers the region

$$-\infty < T < +\infty\,,$$
$$0 < X < +\infty\,,$$
$$-\infty < Y < +\infty\,,$$
$$-\infty < Z < +\infty\,.$$

The Rindler metric is nothing but the Minkowski metric in a chart covering the wedge $|t| < |x|$. Although the Rindler metric does not look flat, a calculation of the Riemann tensor (which vanishes identically) shows that the geometry is indeed flat. The Rindler metric has a coordinate singularity at $X = 0$, which corresponds to the Rindler horizon $t = \pm x$. The metric can be continued analytically beyond the horizon by going back to the original Minkowski coordinates (t,x,y,z).

2.4 Event Horizons

The traditional notion of horizon emerging from the study of static or stationary black holes in General Relativity is that of event horizon. An *event horizon* is *a connected component of the boundary of the causal past of future null infinity* [47, 48, 78, 95]. In the notations of Ref. [95], in which $\mathscr{I}^+$ is future null infinity and $J^-(\mathscr{I}^+)$ is its causal past, i.e., the set of all events which can send lightlike signals to $\mathscr{I}^+$, the event horizon is $\partial\left(J^-(\mathscr{I}^+)\right)$. This definition embodies the most peculiar feature of a black hole, i.e., the horizon is a causal boundary which separates a region from which nothing can come out to reach a distant observer from a region in which signals can be sent out and eventually arrive to this observer. An event horizon is generated by the null geodesics which fail to reach infinity and, therefore (provided that it is smooth) is always a *null hypersurface*.

In black hole research and in astrophysics the concept of event horizon is implicitly taken to define the concept of static or stationary black hole itself. However, since to define and locate an event horizon one must know all the future history of spacetime (one must know all the geodesics which do reach null infinity and, tracing them back, the boundary of the region from which they originate), an event horizon is a *globally* defined concept. To state that an event horizon has formed (which traditionally is understood to mean that a black hole has formed or is about to form) requires knowledge of the spacetime outside our future light cone, which is impossible to achieve (unless, of course, the spacetime is stationary and the black hole has existed forever—then nothing changes and by knowing the state of the world now one knows it forever). It is often said that the event horizon has a teleological nature. It has been shown [6, 19] that, inside a collapsing spherical shell in Vaidya spacetime, an event horizon forms and grows, starting from the centre, and an observer can cross it and be unaware of it even though his or her causal past consists entirely of a portion of flat Minkowski space. In other words, the event horizon "knows" about events belonging to a spacetime region very far away and with no causal connection (a property called "clarvoyance" [18]).

Because of its global nature, an event horizon is not a practical notion to work with, and it is nearly impossible to locate precisely an event horizon in a general dynamical situation. In practice, astrophysical black holes did not exist forever but formed in a highly dynamical process of gravitational collapse. Numerical relativity codes are written to follow a gravitational collapse, the merger of a binary system, or other dynamical situations ending in a black hole, and they crash at some point. It is clearly impossible to follow the evolution of a system all the way to future null infinity. Numerical relativists routinely use marginally trapped surfaces as proxies for event horizons (see, e.g., [16, 29]).

Strictly speaking, an event horizon $\mathscr{H}$ is a tube in spacetime; it is a common abuse of terminology to refer to the intersections of $\mathscr{H}$ with surfaces of constant time (which produce 2-surfaces) as "event horizons". Although improper, this terminology is widespread and extends to the other notions of horizon that we define below.

Example 2.1. In the Schwarzschild geometry

$$ds^2 = -\left(1 - \frac{2M}{R}\right) dt^2 + \frac{dR^2}{1 - \frac{2M}{R}} + R^2 d\Omega_{(2)}^2 \qquad (2.54)$$

the surface $R = 2M$ is a well known event horizon.

Example 2.2. The outer horizon $R_+ = m + \sqrt{m^2 - Q^2}$ of the Reissner-Nordström metric (1.44) is an event horizon. The inner horizon $R_- = m - \sqrt{m^2 - Q^2}$ is not because in the entire region $R < R_+$ there are outgoing radial null geodesics which fail to reach future null infinity and the surface $R = R_-$ is not a boundary of a region with this property.

2.5 Killing Horizons

Remember that a Killing vector field k^a is one that satisfies the Killing equation

$$\nabla_a k_b + \nabla_b k_a = 0 . \qquad (2.55)$$

A Killing vector describes a symmetry of spacetime in a geometric, coordinate-invariant way [38, 48, 95].

A *Killing horizon* $\mathscr{H}$ of the spacetime (M, g_{ab}) is *a null hypersurface which is everywhere tangent to a Killing vector field k^a which becomes null*, $k^c k_c = 0$, on $\mathscr{H}$. This Killing vector field is timelike, $k^c k_c < 0$, in a spacetime region which has $\mathscr{H}$ as boundary. Stationary event horizons in General Relativity are usually Killing horizons for a suitably chosen Killing vector [28]. For example, in the Schwarzschild geometry (1.3), the timelike Killing vector $k^a = (\partial/\partial t)^a$ in the $R > 2M$ region outside the horizon becomes null at $R = 2M$ and spacelike inside (i.e., for $R < 2M$). The event horizon $R = 2M$ is also a Killing horizon. More generally, any event horizon in a locally static spacetime is also a Killing horizon for the Killing vector $k^a = (\partial/\partial t)^a$ associated with the time symmetry. If the spacetime is stationary and asymptotically flat (but not necessarily static), it must be axisymmetric and any event horizon is a Killing horizon for the Killing vector

$$k^a = (\partial/\partial t)^a + \Omega_H (\partial/\partial\varphi)^a , \qquad (2.56)$$

which is a linear combination of the vectors associated with time and rotational symmetries, and where Ω_H is the angular velocity at the horizon (this statement requires the assumption that the Einstein-Maxwell equations hold and some assumption on the matter stress-energy tensor [48, 96]).

The concept of Killing horizon ceases to be useful in spacetimes, or spacetime regions, which are not stationary and do not admit timelike Killing vectors. There

have been attempts to use conformal Killing horizons in spacetimes which are
conformal to the Schwarzschild one ([37, 84], see also [61–63, 85]) but this approach
is a priori very restrictive and, in retrospect, it does not seem to have been very
productive. Instead, the introduction of the Kodama vector, which resembles in some
way a Killing field but is defined in spacetimes without Killing vectors, seems to be
much more useful in defining surface gravities and in the thermodynamics of time-
evolving horizons.

When present, a Killing horizon defines a notion of surface gravity κ_{Killing}, as we
will see in Sect. 2.10.

2.6 Apparent Horizons

A *future apparent horizon*[2] is *the closure of a surface* (usually a 3-surface) *which
is foliated by marginal surfaces* ($\theta_l = 0$). The future apparent horizon is a surface
defined by the conditions on the time slicings [50]

$$\theta_l = 0 \,, \tag{2.57}$$

$$\theta_n < 0 \,, \tag{2.58}$$

where θ_l and θ_n are the expansions of the future-directed outgoing and ingoing null
geodesic congruences, respectively. Equation (2.57) tells us that the future-pointing
outgoing null geodesics momentarily stop propagating outward and, presumably,[3]
turn around at the horizon, while the condition (2.58) distinguishes between black
holes and white holes.

Apparent horizons are defined *quasi-locally*[4] and do not refer to the global causal
structure of spacetime—they don't have the teleological nature of event horizons.
Apparent horizons (and also trapping horizons, see below) depend on the choice
of the foliation of the 3-surface with marginal surfaces [82, 97]. Also the ingoing
and outgoing null geodesics orthogonal to these surfaces and their expansions θ_l
and θ_n depend on the foliation [42]. The expansions θ_l and θ_n are scalars and are,

---

[2]This is not the definition of apparent horizon originally introduced in the book by Hawking and
Ellis [48], which is not easy to work with in practice [21]. It is unfortunate that the term "apparent
horizon" corresponds to different precise definitions in the literature. The definition that we provide
here is more useful than that of [48] in practical (including numerical) applications and is the one
which is adopted in most of the recent literature.

[3]The fact that these null rays "hesitate" ($\theta_l = 0$) does not necessarily imply that they are turning
around and will subsequently propagate inward. One could have, for example, a wormhole throat
at which outgoing null rays "hesitate" and then propagate outward again.

[4]"Quasi-local" refers to a quantity which can be measured by an observer in a finite lifespan
experiment, as opposed to a global quantity which requires the observer to know the entire future
history or causal structure of spacetime which is, of course, physically impossible and would
require an infinite observation time in the non-stationary case.

therefore, independent of the coordinate system chosen, but sometimes a choice of coordinates helps in specifying locally the foliation (for example by choosing spacelike surfaces of constant time coordinate—different time coordinates identify different families of hypersurfaces of constant time), which is a geometric construct and is coordinate-independent. However, congruences of outgoing and ingoing null geodesics orthogonal to these surfaces will, of course, change by changing the foliation while being coordinate-independent. The dependence of apparent horizons on the spacetime slicing is made clear by the fact that non-symmetric slicings of the Schwarzschild spacetime exist which do not admit any apparent horizon [82, 97].

Apparent horizons are, in general, distinct from event horizons: for example, these horizons do not coincide in the Reissner-Nordström black hole (inner horizon) and in the Vaidya spacetime [78]. Also in static black holes which are perturbed, the apparent and the event horizons do not coincide. During the spherical collapse of uncharged matter, an event horizon forms before the apparent horizon does and these two horizons approach each other until they eventually coincide as the final static state is reached [48].

Apparent horizons are always found inside the event horizon provided that the null curvature condition $R_{ab} l^a l^b \geq 0$ ∀ null vector l^a is satisfied [48]. This requirement coincides with the null energy condition $T_{ab} l^a l^b \geq 0$ ∀ null vector l^a if the Einstein equations are imposed, and in this case it is believed to be a "reasonable" condition on physical matter. But the Hawking radiation produced by horizons violates the weak and the null energy conditions [92]. Also a simple scalar field non-minimally coupled to the curvature can violate the energy conditions [17] and the null curvature condition is easily violated in alternative theories of gravity (for example, Brans-Dicke [26] and scalar-tensor [20, 74, 94] theories) and the apparent horizon is reported to lie outside of the event horizon during spherical collapse in Brans-Dicke gravity, although it eventually settles inside the event horizon when the static Schwarzschild configuration is achieved [81]. One should not be too attached to the notion of event horizon and should perhaps look at other notions of horizon as more realistic even though they depend on the spacetime slicing (apparent, trapping, and dynamical horizons have this dependence).

2.7 Trapping Horizons

A *future outer trapping horizon* (FOTH) is *the closure of a surface* (usually a 3-surface) *foliated by marginal surfaces such that on its 2-dimensional "time slicings" the conditions* ([50], see also [73] and references therein)

$$\theta_l = 0 \,, \tag{2.59}$$

$$\theta_n < 0 \,, \tag{2.60}$$

$$\mathscr{L}_n \theta_l = n^a \nabla_a \theta_l < 0 \,, \tag{2.61}$$

are satisfied, where θ_l and θ_n are the expansions of the future-directed outgoing and ingoing null geodesic congruences, respectively. The condition (2.61) is introduced to distinguish between inner and outer horizons (e.g., in the non-extremal Reissner-Nordström solution) and also distinguishes between apparent horizons and trapping horizons (it is not imposed for apparent horizons but it is required for trapping ones).

One obtains the definition of *past inner trapping horizon* (PITH) by exchanging l^a with n^a while reversing the signs of the inequalities,

$$\theta_n = 0, \tag{2.62}$$

$$\theta_l > 0, \tag{2.63}$$

$$\mathscr{L}_l \theta_n = l^a \nabla_a \theta_n > 0. \tag{2.64}$$

The past inner trapping horizon characterizes a white hole or a cosmological horizon.

If we don't exchange l^a and n^a, we can say that, in general, a *trapping horizon* (TH) satisfies the definition requirements [21]

$$\theta_l = 0, \tag{2.65}$$

$$\theta_n \neq 0, \tag{2.66}$$

$$\mathscr{L}_n \theta_l \neq 0, \tag{2.67}$$

and is

- *Future* if $\theta_n < 0$,
- *Past* if $\theta_n > 0$,
- *Outer* if $n^a \nabla_a \theta_l < 0$,
- *Inner* if $n^a \nabla_a \theta_l > 0$.

For black holes, the trapping horizon has been associated with thermodynamics, and it has even been claimed that it is the trapping horizon area and not the area of the event horizon which should be associated with entropy in black hole thermodynamics [31, 45, 56, 68]. This claim, however, is the subject of controversy [32, 70, 83]. The Parikh-Wilczek "tunneling" approach [75] predicts Hawking radiation also for apparent and for trapping horizons, not only for event horizons [30, 35, 54, 58, 67, 71, 93] but also this aspect is not entirely free of controversy [15].

Trapping horizons do not, in general, coincide with event horizons. Dramatic examples are spacetimes which possess trapping horizons but not event horizons [53, 80]. The difference between the areas of the trapping and the event horizon in particular spacetimes have been studied in Ref. [69].

Example 2.3. In the Reissner-Nordström spacetime with the natural spherically symmetric foliation, the event horizon $r = r_+$ is a future outer trapping horizon (FOTH), the inner Cauchy horizon $r = r_-$ is a future inner trapping horizon (FITH), while the white hole horizons are past trapping horizons (PTHs).

2.8   Isolated and Dynamical Horizons

In the realm of black holes, isolated horizons correspond to isolated systems in thermal equilibrium not interacting with their surroundings, which could instead be dynamical. The concept of isolated horizon has been introduced in the context of loop quantum gravity [3, 7–11, 13, 41]. In general, this horizon construct is too restrictive when one wants to allow mass-energy to cross the "horizon" in whatever direction.

A *weakly isolated horizon* is *a null surface $\mathcal{H}$ with null normal l^a such that* $\theta_l = 0$, $-T_{ab}l^a$ *is a future-oriented and causal vector, and* $\mathcal{L}_l \left(n_b \nabla_a l^b \right) = 0$. In this context l^a is a Killing vector of the *intrinsic* geometry on $\mathcal{H}$, without reference to the surroundings, and can be used to define a "completely local Killing horizon" in the absence of energy fluxes across $\mathcal{H}$. The vector field l^a generates a congruence of null geodesics on $\mathcal{H}$, which can be employed to define a surface gravity κ using the non-affinely parametrized geodesic equation

$$l^a \nabla_a l^b = \kappa \, l^b \,, \tag{2.68}$$

which also yields

$$\kappa = -n_b l^a \nabla_a l^b \,, \tag{2.69}$$

where $n_b l^b = -1$. The surface gravity κ defined in this way is constant on the weakly isolated horizon $\mathcal{H}$, corresponding to the zeroth law of thermodynamics. The vector field n^a is not unique, hence this notion of surface gravity is not defined uniquely.

An Hamiltonian analysis of the phase space of isolated horizons identifies boundary terms with energies of these boundaries and produces a first law of thermodynamics for isolated horizons with rotational symmetry, i.e.,

$$\delta H_{\mathcal{H}} = \frac{\kappa}{8\pi} \, \delta \mathcal{A} + \Omega_{\mathcal{H}} \delta J \,. \tag{2.70}$$

Here J is the angular momentum, $H_{\mathcal{H}}$ is the Hamiltonian, $\mathcal{A}$ is the area of the 2-dimensional cross-sections of $\mathcal{H}$, and $\Omega_{\mathcal{H}}$ is the angular velocity of the horizon.

Next, a *dynamical horizon* [6] is *a spacelike marginally trapped tube* (MTT). This definition allows energy flows to cross the dynamical horizon. A set of flux laws describing the related changes of the area of the dynamical horizons are available [6]. An apparent horizon which is everywhere spacelike is a dynamical horizon, but an apparent horizon is not required to be spacelike.[5] Being spacelike, dynamical horizons can be crossed only in one direction by causal curves, while this is not the case for apparent horizons which can be partially or entirely non-spacelike.

---

[5]An apparent horizon which is everywhere timelike is called a *timelike membrane* [4, 5, 24].

To end this string of definitions, *slowly evolving horizons* have also been defined and examined [21–23, 59]: they are "almost isolated" FOTHs and they are meant to characterize black hole horizons which evolve very slowly in time. This slow evolution is expected in many astrophysical processes but not, for example, in the final stages of collapse or evaporation. Slowly evolving horizons are analogous to thermodynamic systems in a regime near equilibrium.

2.9 Kodama Vector and Surface Gravity

Various notions of surface gravity associated with horizons have been introduced in the literature. In static and stationary situations a timelike Killing vector field is present outside the horizon and becomes null on it, and these notions of surface gravity coincide and are well known from the study of the Kerr-Newman black holes of General Relativity. In dynamical situations, however, there is no timelike Killing vector and the various notions of surface gravity encountered in the literature turn out to be inequivalent. In spherical symmetry, the Kodama vector mimics the properties of a Killing vector and gives rise to a conserved current and a surface gravity.

The Kodama vector [60] generalizes the notion of Killing vector field to spacetimes which do not admit one and has been used as a substitute of a Killing vector in the thermodynamics of time-dependent horizons. The Kodama vector is defined only *in spherical symmetry*.[6] Write the spacetime metric as

$$ds^2 = h_{ab}dx^a dx^b + R^2 d\Omega_{(2)}^2 \,, \tag{2.71}$$

where $a, b = 0, 1$ and R is the areal radius, and let ϵ_{ab} be the volume form of the 2-metric h_{ab} [95]; then the *Kodama vector* is defined as [60]

$$K^a \equiv \epsilon^{ab}\nabla_b R \tag{2.72}$$

with $K^\theta = K^\varphi = 0$. The Kodama vector lies in the 2-dimensional (t, R) surface (where t is the time coordinate) orthogonal to the 2-spheres of symmetry.[7] In a static spacetime the Kodama vector is parallel (in general, not equal) to the timelike Killing vector. In the region in which it is timelike, the Kodama vector defines a class of preferred observers with 4-velocity $u^a \equiv K^a / \sqrt{|K^c K_c|}$.

---

[6]Reference [88] attempts to generalize the Kodama vector to non-spherically symmetric spacetimes.

[7]Moreover, $K^a \nabla_a R = \epsilon^{ab}\nabla_a R \nabla_b R = 0$ because ϵ^{ab} is antisymmetric and $\nabla_a R \nabla_b R$ is symmetric.

It can be proved ([60], see [1] for a simplified proof) that the Kodama vector has zero divergence

$$\nabla_a K^a = 0;$$ (2.73)

this property has the consequence that the *Kodama energy current*

$$J^a \equiv G^{ab} K_b$$ (2.74)

(where G_{ab} is the Einstein tensor) is covariantly conserved, $\nabla^a J_a = 0$ even if there is no timelike Killing vector. This surprising property is sometimes called the "Kodama miracle" [1, 60].

If the spherically symmetric metric is written in Schwarzschild-like coordinates,

$$ds^2 = -A(t, R)\, dt^2 + B(t, R)\, dR^2 + R^2 d\Omega_{(2)}^2,$$ (2.75)

then the Kodama vector takes on the simple form (e.g., [60, 79])

$$K^a = \frac{-1}{\sqrt{AB}} \left(\frac{\partial}{\partial t} \right)^a.$$ (2.76)

Proof. Denoting by h the determinant of the 2-metric h_{ab}, we have $\sqrt{|h|} = \sqrt{AB}$ and the volume form of h_{ab} is

$$\epsilon_{ab} = \sqrt{AB} \left(\nabla_a t \nabla_b R - \nabla_a R \nabla_b t \right)$$

so that

$$\begin{aligned}
\epsilon^{ab} &= g^{ac} g^{bd} \epsilon_{cd} = g^{ac} g^{bd} \sqrt{AB} \left(\delta_{c0}\delta_{d1} - \delta_{d0}\delta_{c1} \right) \\
&= \sqrt{AB} \left(g^{a0} g^{b1} - g^{b0} g^{a1} \right).
\end{aligned}$$

Then the only non-vanishing components of ϵ^{ab} are $\epsilon^{01} = -\epsilon^{10} = (AB)^{-1/2}$ and the Kodama vector has components

$$K^\mu = \epsilon^{\mu\nu} \nabla_\nu R = \epsilon^{\mu 1} = \sqrt{AB}\, \delta^{\mu 0} g^{00} g^{11} = \frac{-\delta^{\mu 0}}{\sqrt{AB}} = \frac{-1}{\sqrt{AB}} \left(\frac{\partial}{\partial t} \right)^\mu.$$

$\square$

It is shown in [51] that the Noether charge associated with the Kodama conserved current is the Misner-Sharp-Hernandez mass [55, 64]. This notion of mass energy, too, is defined only in spherical symmetry.

The Kodama vector was used in Ref. [79] (see also [33]) to simplify the evolution equations and the initial value problem for the coupled Einstein-Klein-Gordon equations with spherical symmetry.

2.10   Surface Gravities

Surface gravity is defined classically in terms of geometric properties of the metric tensor. However, the same quantity appears in black hole thermodynamics as the proportionality factor between the variation of the black hole mass (acting as the internal energy) dM and the variation of the event horizon area (proportional to the entropy) dA. Since it is not established at present which definition of black hole mass is appropriate in non-trivial backgrounds (different definitions are possible, see the review [86]), it is obvious that also the definition of surface gravity will be subject to the same ambiguities. Moreover, surface gravity appears also semiclassically since, up to a numerical constant, it coincides with the Hawking temperature of a black hole.

The traditional definition of surface gravity is given on a Killing horizon for stationary spacetimes [95]. However, given that Killing horizons are not defined in general non-stationary situations in which one considers quasi-local horizons instead of event/Killing horizons, a different concept of surface gravity is necessary in these cases. There are several definitions of surface gravity in the literature, and they are inequivalent. The recurrent definitions are reviewed in Ref. [72] and are briefly recalled here.

2.10.1   Killing Horizon Surface Gravity

As already remarked, a Killing horizon defines a notion of surface gravity κ_{Killing} as follows [95]: on the Killing horizon the Killing vector k^a satisfies

$$k^a \nabla_a k^b \equiv \kappa_{\text{Killing}} k^b , \tag{2.77}$$

so κ_{Killing} measures the failure of the geodesic Killing vector k^a to be affinely-parametrized (the "inaffinity") on the Killing horizon.

Proof. We need to prove [72] that, on the Killing horizon, $k^a \nabla_a k^b$ is proportional to k^b. First, note that the equation defining the Killing horizon is $k^c k_c = 0$, hence $\nabla_b(k^c k_c) = 2k^c \nabla_b k_c$ is orthogonal to the horizon. But since k^a is also normal to the horizon (because k^a is null there) and it generates this horizon, then $k^c \nabla_b k_c \propto k_b$. Using the Killing equation $\nabla_a k_b + \nabla_b k_a = 0$, one obtains

$$k^c \nabla_b k_c = -k^c \nabla_c k_b \propto k_b ,$$

so $k^c \nabla_c k_b \propto k_b$, or $k^c \nabla_c k_b = \kappa_{\text{Killing}} k_b$. $\square$

Another property of the Killing surface gravity is (cf. Ref. [95], p. 332)

$$\kappa^2_{\text{Killing}} = -\frac{1}{2} \left(\nabla^a k^b \right) \left(\nabla_a k_b \right) . \tag{2.78}$$

In static spacetimes (and only in those!) one can interpret the surface gravity κ_{Killing} as the limiting force required at spatial infinity to hold in place a unit test mass just above the event horizon by means of an infinitely long massless string [95]. This interpretation shows the non-local nature of the notion of Killing surface gravity.

Since the Killing equation $\nabla_a k_b + \nabla_b k_a = 0$ determines the Killing vector k^a only up to an overall normalization, there is freedom to rescale k^a and the value of the surface gravity depends on the non-affine parametrization chosen for k^a. However, in stationary situations one has the freedom of imposing that $k^c k_c = -1$ at spatial infinity.

Example 2.4. For the Kerr-Newman black hole (1.70)–(1.74), the Killing surface gravity is [95]

$$\kappa_{\text{Killing}} = \frac{\sqrt{M^2 - a^2 - Q^2}}{2M \left(M + \sqrt{M^2 - a^2 - Q^2} \right) - Q^2} . \tag{2.79}$$

The Killing surface gravity can be generalized to any event horizon that is not a Killing horizon by replacing the Killing vector k^a with the null generator of the event horizon [72].

2.10.2 Surface Gravity of Marginally Trapped Surfaces

Consider the outgoing and ingoing null normals l^a and n^a to a marginally trapped (spacelike compact 2-dimensional) surface, and assume that the expansion of l^a vanishes, with l^a and n^a normalized so that $l^c n_c = -1$. l^a is not a horizon generator in general, but it is still a non-affinely parametrized geodesic vector on the trapping horizon. This fact allows one to introduce a surface gravity κ by

$$l^a \nabla_a l^b \equiv \kappa \, l^b \tag{2.80}$$

or

$$\kappa = -n^b l^a \nabla_a l_b . \tag{2.81}$$

The value of this surface gravity depends on the parametrization of l^a, for which various proposals have been advanced. By writing l^a as the tangent to a null curve $x^\mu(\lambda)$ with parameter λ, a parameter change (which depends on the spacetime point) $\lambda \to \lambda'$ forces the components of l^a to change as

$$l^\mu = \frac{dx^\mu}{d\lambda} \to l^{\mu'} = \frac{dx^\mu}{d\lambda'} = l^\mu \frac{d\lambda}{d\lambda'} = \Omega(x) \, l^\mu \tag{2.82}$$

so that

$$l^{\nu'} \nabla_{\nu'} l^{\mu'} = \kappa' l^{\mu'} \,,$$

$$\Omega l^\nu \nabla_\nu \left(\Omega l^\mu \right) = \kappa' \Omega l^\mu \,,$$

$$\Omega \kappa \, l^\mu + \left(l^\nu \nabla_\nu \Omega \right) l^\mu = \kappa' l^\mu \,, \tag{2.83}$$

and

$$\kappa \to \kappa' = \Omega \, \kappa + l^\nu \nabla_\nu \Omega \,. \tag{2.84}$$

The *Hayward proposal for spherical symmetry* [52] uses the Kodama vector K^a which is always available in spherical symmetry [60]. This future-directed vector satisfies

$$\nabla_b \left(K_a T^{ab} \right) = 0 \,. \tag{2.85}$$

The Hayward notion of surface gravity κ_{Kodama} for a trapping horizon is given by

$$\frac{1}{2} g^{ab} K^c \left(\nabla_c K_a - \nabla_a K_c \right) = \kappa_{\text{Kodama}} K^b \,. \tag{2.86}$$

This definition is unique because of the uniqueness of the Kodama vector. The surface gravity κ_{Kodama} agrees with the surface gravity on the horizon of a Reissner-Nordström black hole but not with other dynamical surface gravity constructs. An expression equivalent to (2.86) is [52]

$$\kappa_{\text{Kodama}} = \frac{1}{2} \Box_{(h)} R = \frac{1}{2\sqrt{-h}} \partial_\mu \left(\sqrt{-h} \, h^{\mu\nu} \partial_\nu R \right) \,, \tag{2.87}$$

where h is the determinant of the metric h_{ab} in the 2-space orthogonal to the 2-spheres of symmetry. The Hamilton-Jacobi approach (a variant of the Parikh-Wilczek method [75]) to study the Hawking radiation of time-dependent horizons, leads naturally to the Kodama-Hayward definition of surface gravity [36] (see Ref. [90] for a review of tunneling methods).

2.10.3 Fodor et al. Surface Gravity

The proposal by Fodor et al. [43] applies to spherically symmetric and asymptotically flat spacetimes and is based on the normalization $l^a t_a = -1$ of the ingoing null normal n^a, where t^a is the asymptotic time-translational Killing vector at spatial infinity. The curve with tangent n^a is affinely-parametrized everywhere. By requiring that $l^c n_c = -1$, this choice fixes the parametrization of l^a and one obtains [43]

$$\kappa_{\text{Fodor}} = -n^b l^a \nabla_a l_b . \tag{2.88}$$

2.10.4   Isolated Horizon Surface Gravity

This proposal, due to Ashtekar, Beetle, and Fairhurst [8] applies to an isolated horizon. The null normal n^a is normalized in such a way that its expansion agrees with that of the Reissner-Nordström black hole and with $l^a n_a = -1$. This normalization identifies a unique surface gravity as a function of the horizon parameters. This notion of surface gravity appears to be quite limited, for example it cannot be extended to the Einstein-Yang-Mills case [12, 72].

2.10.5   Proposal for Slowly Evolving Horizons

The notion of slowly evolving horizons [22] extends the previous proposal for surface gravity. On the isolated horizon the normal is $\tau^a = Bl^a + Cn^a$, with B and C scalar fields defined there, which weigh the contributions of l^a and n^a (for an isolated horizon it is $B = 1$, $C = 0$). The slowly evolving horizon surface gravity is

$$\kappa_{\text{SE}} \equiv -Bn^a l^b \nabla_b l_a - Cl^a n^b \nabla_b n_a . \tag{2.89}$$

2.10.6   Other Proposals

There are other proposals for surface gravity in the literature, including Hayward's trapping gravity [50]

$$\kappa_{\text{trapping}} \equiv \frac{1}{2} \sqrt{-n^a \nabla_a \theta_l} \tag{2.90}$$

and the Mukohyama-Hayward proposal ([66], see also [72]).

All these definitions have been computed, for a general spherically symmetric metric in Eddington-Finkelstein coordinates and in terms of the Misner-Sharp-Hernandez mass [55, 64], in Ref. [72]. A critical comparison of these definitions for black holes in spherical symmetry and using Painlevé-Gullstrand coordinates is contained in Ref. [77].

2.11 Spherical Symmetry

The assumption of spherical symmetry greatly simplifies the study of horizons and the solution of the field equations. While exact spherical symmetry is unrealistic for astrophysical black holes (which rotate and may be distorted by other bodies and by magnetic fields) and for realistic universes perturbed by non-spherical inhomogeneities, it is an important assumption for the fundamental theory and it may also be a realistic approximation in certain situations, especially in cosmology.

In spherical symmetry, the apparent horizons (existence, location, dynamics, surface gravity, etc.) can be studied by using the Misner-Sharp-Hernandez mass M_{MSH} [55, 64], which coincides with the Hawking-Hayward quasi-local energy [46, 49] for spherical spacetimes. In General Relativity the Misner-Sharp-Hernandez mass is defined only for spherically symmetric spacetimes. In terms of the areal radius R and angular coordinates (θ, φ), a spherically symmetric line element is written as

$$ds^2 = h_{ab}dx^a dx^b + R^2 d\Omega^2_{(2)}, \qquad (2.91)$$

where $a, b = 1, 2$. The Misner-Sharp-Hernandez mass M_{MSH} is defined by [55, 64]

$$1 - \frac{2M_{\mathrm{MSH}}}{R} \equiv \nabla^c R \nabla_c R \qquad (2.92)$$

or by[8]

$$M_{\mathrm{MSH}} = \frac{R}{2} \left(1 - h^{ab}\nabla_a R \nabla_b R\right). \qquad (2.93)$$

Note that the Misner-Sharp-Hernandez mass is an invariant quantity of the 2-space normal to the 2-spheres of symmetry.

Horizons in the presence of spherical symmetry are discussed in a clear and elegant way using the Nielsen-Visser formalism ([73], see also Ref. [71]). They consider the most general spherically symmetric metric with a spherically symmetric spacetime slicing, which assumes the form

$$ds^2 = -e^{-2\phi(t,R)}\left[1 - \frac{2M(t,R)}{R}\right]dt^2 + \frac{dR^2}{1 - \frac{2M(t,R)}{R}} + R^2 d\Omega^2_{(2)} \qquad (2.94)$$

---

[8]In $(n+1)$ spacetime dimensions, the Misner-Sharp-Hernandez mass is $M_{\mathrm{MSH}} = \frac{n(n-1)}{16\pi}R^{n-2}V_n\left(1 - h^{ab}\nabla_a R\nabla_b R\right)$, where the line element is $ds^2 = h_{ab}dx^a dx^b + R^2 d\Omega^2_{(n-1)}$ $(a, b = 1, 2)$ and $V_n = \frac{\pi^{n/2}}{\Gamma\left(\frac{n}{2}+1\right)}$ is the volume of the $(n-1)$-dimensional unit ball [14].

in Schwarzschild-like coordinates. A posteriori, $M(t, R)$ turns out to be the Misner-Sharp-Hernandez mass. This form is ultimately inspired by the Morris-Thorne wormhole metric [65], it compromises between the latter and the gauge (2.75), and is particularly convenient in the study of both static and time-varying black holes [73, 91]. It is not assumed that the spacetime is stationary or asymptotically flat.

The line element (2.94) can be rewritten in *Painlevé-Gullstrand coordinates* as (see Appendix A.1)

$$ds^2 = -\frac{e^{-2\phi}}{(\partial\tau/\partial t)^2}\left(1 - \frac{2M}{R}\right)d\tau^2 + \frac{2e^{-\phi}}{\partial\tau/\partial t}\sqrt{\frac{2M}{R}}\, d\tau dR + dR^2 + R^2 d\Omega_{(2)}^2 , \quad (2.95)$$

where $\phi(\tau, R)$ and $M(\tau, R)$ are implicit functions of (τ, R) and the hypersurfaces[9] $\tau = $ constant are flat (setting $d\tau = 0$ gives $ds_{(3)}^2 = dR^2 + R^2 d\Omega_{(2)}^2$, the 3-dimensional Euclidean metric in spherical coordinates). By defining the implicit functions of (τ, R) [73]

$$c(\tau, R) \equiv \frac{e^{-\phi(t,R)}}{(\partial\tau/\partial t)} , \quad (2.96)$$

$$v(\tau, R) \equiv \sqrt{\frac{2M(t,R)}{R}}\,\frac{e^{-\phi(t,R)}}{\partial\tau/\partial t} = c\sqrt{\frac{2M}{R}} , \quad (2.97)$$

the line element is rewritten as

$$ds^2 = -\left[c^2(\tau, R) - v^2(\tau, R)\right]d\tau^2 + 2v(\tau, R)\,d\tau dR + dR^2 + R^2 d\Omega_{(2)}^2 . \quad (2.98)$$

We now list a number of results obtained in [73] which are useful for practical computations in spherical symmetry.

The outgoing radial null geodesic congruence has tangent field

$$l^\mu = \frac{1}{c(\tau, R)}\left(1, c(\tau, R) - v(\tau, R), 0, 0\right), \quad (2.99)$$

in Painlevé-Gullstrand coordinates $(\tau, R, \theta, \varphi)$, while the ingoing radial null geodesics have tangent field

$$n^\mu = \frac{1}{c(\tau, R)}\left(1, -c(\tau, R) - v(\tau, R), 0, 0\right), \quad (2.100)$$

---

[9]The hypersurfaces $\tau = $ constant are obviously spacelike since τ is a timelike coordinate. To be explicit, the normal $N_a = \nabla_a \tau$ to a surface $\tau = $ constant has norm squared $N_a N^a = g_H^{00} = -e^{2\phi_H} < 0$ and N^a is timelike.

with the normalization [73]

$$g_{ab}l^a n^b = -2 \,. \tag{2.101}$$

The expansions of these radial null geodesic congruences are computed as

$$\theta_l = \frac{2}{R}\left(1 - \sqrt{\frac{2M}{R}}\right), \tag{2.102}$$

$$\theta_n = -\frac{2}{R}\left(1 + \sqrt{\frac{2M}{R}}\right). \tag{2.103}$$

A 2-sphere of symmetry of radius R is [51, 73, 78]

- *Trapped* if $R < 2M$,
- *Marginal* if $R = 2M$,
- *Untrapped* if $R > 2M$.

The *apparent horizon* is the boundary between trapped and untrapped surfaces and corresponds to $\theta_l = 0$ and $\theta_n < 0$ and is, therefore, given by

$$\frac{2M\,(\tau, R_{AH})}{R_{AH}(\tau)} = 1 \iff \nabla^c R \nabla_c R \,|_{AH} = 0 \iff g^{RR}\,|_{AH} = 0 \,, \tag{2.104}$$

where the last equation holds in both Painlevé-Gullstrand coordinates and in the gauge (2.94) and is obtained by using the inverse of the metric (2.98)

$$(g^{\mu\nu}) = \frac{1}{c^2}\begin{pmatrix} -1 & v & 0 & 0 \\ v & c^2 - v^2 & 0 & 0 \\ 0 & 0 & R^2 & 0 \\ 0 & 0 & 0 & R^2\sin^2\theta \end{pmatrix}. \tag{2.105}$$

The condition $g^{RR} = 0$ is a very convenient and practical recipe to locate the apparent horizons in spherical symmetry when the areal radius R is used as a coordinate. Sometimes it is convenient to perform a coordinate transformation to this radial coordinate to rewrite the line element using R explicitly.

The gradient of the areal radius R and the normal $n_a = \nabla_a R$ to the surfaces $R = $ constant become null at the apparent horizon. It is clear that this recipe resembles the change in causal character of the Schwarzschild radial coordinate (which is also an areal radius) on the Schwarzschild event horizon. In general, however, the apparent horizon is not a null surface like the event horizon of the Schwarzschild black hole (see the examples below).

Nielsen and Visser provide also the derivative

$$\mathscr{L}_n \theta_l \mid_{\text{AH}} = n^a \nabla_a \left[\frac{2}{R} \left(1 - \sqrt{\frac{2M}{R}} \right) \right]_{\text{AH}} = -\frac{2 \left(1 - 2M'_{\text{AH}} \right)}{R^2_{\text{AH}}} \left(1 + \frac{\dot{R}_{\text{AH}}}{2c_{\text{AH}}} \right),$$

(2.106)

with a prime and an overdot denoting partial differentiation with respect to R and τ, respectively, and with the subscript AH denoting quantities evaluated on the apparent horizon. It is noted in Refs. [71, 73] that $1 - 2M'_{\text{AH}} > 0$ is required for the horizon to be outer in a spacetime with regular asymptotic region, hence *the condition for the apparent horizon to be also a trapping horizon is*

$$\dot{R}_{\text{AH}} > -2c_{\text{AH}}.$$

(2.107)

Equation (2.103) tells us that along ingoing radial null geodesics with tangent n^a it is $\dot{R} = -2e^{-\phi}$, hence the apparent horizon at $R_{\text{AH}} = 2M_{\text{AH}}$ is a trapping horizon if it is outer ($2M'_{\text{AH}} < 1$) and does not move inward faster than the ingoing radial null geodesics (in which case the latter would not be trapped).

If matter satisfies the null energy condition, and assuming the Einstein equations, the area of the apparent horizon cannot decrease. Various energy fluxes across the apparent horizon are also introduced and computed in [73]. The Nielsen-Visser surface gravity at the horizon is computed using $l^b \nabla_b l^a = \kappa_l l^a$, obtaining [73]

$$\kappa_l(\tau) = \frac{1 - 2M' \left(\tau, R_{\text{H}}(\tau) \right)}{2R_{\text{H}}(\tau)}.$$

(2.108)

An *extremal horizon* is one which has vanishing surface gravity,

$$1 - 2M' \left(\tau, R_{\text{H}}(\tau) \right) = 0.$$

(2.109)

The fact that the Misner-Sharp-Hernandez mass (which coincides with the Hawking-Hayward quasi-local energy in spherical symmetry [46, 49]) can be employed to locate apparent horizons in spherical symmetry makes it clear once again that the apparent horizon is a quasi-local concept and is independent of the global causal structure. However, it is not a completely local concept.

Using the Misner-Sharp-Hernandez mass it is easy to see when there are an inner and an outer horizon. When this situation happens, it is $M'_{\text{MSH}} > 1/2$ at the inner horizon and $M'_{\text{MSH}} < 1/2$ at the outer horizon. Graphically, this means that (at a given time τ, or at all times if the metric is stationary) at the intersections between the graph of the function $M_{\text{MSH}}(R)$ and the line $M_{\text{MSH}} = R/2$ (which are apparent horizons), the curve $M_{\text{MSH}}(R)$ is steeper [respectively, less steep] than this line. If $M_{\text{MSH}}(R)$ rises faster than $R/2$, in an asymptotically flat spacetime in which $M_{\text{MSH}}(R)$ eventually asymptotes to a constant as $R \to +\infty$, a continuous M_{MSH} will have to cross the line $R = M_{\text{MSH}}/2$ again, and there will be an outer horizon [71].

2.12 Rindler Horizons Revisited

Armed with the notions of Killing horizon and Killing surface gravity, let us revisit now the Rindler horizon of a uniformly accelerated observer in Minkowski spacetime. The Killing vector field of this space associated with Lorentz boosts in the x-direction has components

$$\xi^\alpha = x \left(\frac{\partial}{\partial t} \right)^\alpha + t \left(\frac{\partial}{\partial x} \right)^\alpha , \tag{2.110}$$

or $\xi^\mu = (x, t, 0, 0)$ in Cartesian coordinates (t, x, y, z). Since $\xi^c \xi_c = -x^2 + t^2$, this vector is timelike if $|x| > |t|$, null on the light cone $t = \pm x$, and spacelike if $|x| < |t|$. Therefore, *there is a Killing horizon at $t = \pm x$.*

Along the worldline (2.40) of a uniformly accelerated observer it is

$$\xi^c \xi_c = -x^2 + t^2 = -\frac{1}{a^2} \tag{2.111}$$

and, therefore, we are led to normalize the Killing vector according to

$$k^\alpha \equiv a\xi^\alpha = ax \left(\frac{\partial}{\partial t} \right)^\alpha + at \left(\frac{\partial}{\partial x} \right)^\alpha , \tag{2.112}$$

or $k^\mu = (ax, at, 0, 0)$ in Cartesian coordinates with norm squared

$$k^a k_a = a^2 \left(-x^2 + t^2 \right) .$$

Along the worldline (2.40) of a uniformly accelerated observer, k^a is normalized to -1.

The vanishing of the norm of the Killing vector k^a describes a Killing horizon for the uniformly accelerated observer; the entire Minkowski space can be threaded by observers and their associated Rindler horizons.[10] Now, we can associate a Killing surface gravity κ_{Killing} to this Rindler-Killing horizon as specified by Eq. (2.77). We have

$$k^\alpha \nabla_\alpha k^\beta = k^0 \partial_t k^\beta + k^1 \partial_x k^\beta = ax \, \partial_t k^\beta + at \, \partial_x k^\beta$$

and

$$k^\nu \nabla_\nu k^\mu = \left(a^2 t, a^2 x, 0, 0 \right) . \tag{2.113}$$

---

[10]This fact is essential in Jacobson's thermodynamics of spacetime formalism [40, 57].

On the Rindler-Killing horizon $t = \pm x$, it is $k^\nu \nabla_\nu k^\mu \,|_{\text{RH}} = (\pm a^2 x, a^2 x, 0, 0)$ and Eqs. (2.77) and (2.113) give

$$\left(\pm a^2 x, a^2 x, 0, 0 \right) = \kappa_{\text{Killing}} \left(ax, \pm ax, 0, 0 \right) . \tag{2.114}$$

In order for the Killing vector (2.112) to be future-oriented on the horizon $t = \pm x$, one needs to choose the positive sign in $k^\mu \,|_{\text{RH}} = (\pm at, at, 0, 0)$ and in Eq. (2.114), obtaining the *Killing surface gravity of the Rindler horizon*

$$\kappa_{\text{Killing}} = a , \tag{2.115}$$

which coincides with the uniform acceleration of the Rindler observer. The Unruh temperature (2.44) can then be written as

$$k_B T = \left(\frac{\hbar}{c} \right) \frac{\kappa_{\text{Killing}}}{2\pi} . \tag{2.116}$$

2.13　Conclusions

We are now aware of the various notions of horizon and of surface gravity in the literature. We have studied, in particular, the situation of spherical symmetry which will accompany us for the rest of these lectures. Our simplified exposition is not comprehensive and can certainly be made more rigorous: the reader can find more detailed and technically more satisfactory treatments in the references provided, but those would break the flow of our discussion. We are now going to apply the theory of horizons discussed in this chapter to specific solutions of the Einstein equations (and, later, of the field equations of alternative theories of gravity), in the presence of spherical symmetry. The simplest situation that comes to mind is that of horizons in spatially homogeneous and isotropic cosmologies, and its study is our next step. In spite of its simplicity, this situation is not entirely trivial.

Problems

2.1. The tangent field l^a to a null geodesic congruence is rescaled according to $l^a \to \alpha \, l^a$, where α is a positive constant. How does the expansion θ_l change? How does it change if the null vector n^a used to define the 2-metric h_{ab} orthogonal to l^a is simultaneously rescaled as $n^a \to \alpha^{-1} \, n^a$ preserving the normalization $l^c n_c = -1$?

2.2. Find the outgoing and ingoing radial null geodesic congruences l^a and n^a of the Reissner-Nordström metric and normalize them so that $l^a n_a = -1$.

2.3. Derive Eqs. (2.42) and (2.43).

2.4. Compute θ_l, θ_n, and $n^a \nabla_a \theta_l$ for the non-extremal Reissner-Nordström solution.

2.5. Verify that (2.110) is a Killing vector of Minkowski spacetime.

2.6. Use Eq. (2.78) to obtain directly Eq. (2.115).

2.7. Verify that the time and spatial translations, spatial rotations, and Lorentz boosts given by

$$\xi^a_{(t)} = \left(\frac{\partial}{\partial t}\right)^a,$$

$$\xi^a_{(i)} = \left(\frac{\partial}{\partial x^i}\right)^a \quad (i = 1, 2, 3),$$

$$L^a_{(i)} = \left(\epsilon_{ij}{}^k x^j \frac{\partial}{\partial x^k}\right)^a \quad (i, j, k = 1, 2, 3 \text{ cyclic}),$$

$$\xi^a_{(i)} = x^i \left(\frac{\partial}{\partial t}\right)^a + t \left(\frac{\partial}{\partial x^i}\right)^a \quad (i = 1, 2, 3),$$

respectively, are Killing fields of the Minkowski metric (in fact they are *all* the Killing vector fields of this metric).

2.8. Compute the Misner-Sharp-Hernandez mass of a sphere of constant areal radius for the Reissner-Nordström metric (1.44). Does it reduce to a familiar result in the $Q \to 0$ limit to Schwarzschild? Check[11] that the extremal Reissner-Nordström black hole corresponds to $1 - 2M'_{\text{MSH}}(R_{\text{H}}) = 0$.

2.9. Given the Rindler metric in spherical coordinates

$$ds^2 = -a^2 R^2 \cos^2 \theta \cos^2 \varphi \, dT^2 + dR^2 + R^2 d\Omega^2_{(2)},$$

compute the tangents $l^a_{(\pm)}$ to the congruences of ingoing and outgoing radial null geodesics and their expansions $\theta_{(\pm)}$. Show that there are no apparent horizons in the spacetime region covered by the Rindler chart.

References

1. Abreu, G., Visser, M.: Kodama time: geometrically preferred foliations of spherically symmetric spacetimes. Phys. Rev. D **82**, 044027 (2010)
2. Andersson, L., Mars, M., Simon, W.: Local existence of dynamical and trapping horizons. Phys. Rev. Lett. **95**, 11102 (2005)

---

[11]Cf. Refs. [49, 73].

3. Ashtekar, A., Corichi, A.: Laws governing isolated horizons: inclusion of dilaton couplings. Class. Quantum Grav. **17**, 1317 (2000)
4. Ashtekar, A., Galloway, G.J.: Some uniqueness results for dynamical horizons. Adv. Theor. Math. Phys. **9**, 1 (2005)
5. Ashtekar, A., Krishnan, B.: Dynamical horizons and their properties. Phys. Rev. D **68**, 104030 (2003)
6. Ashtekar, A., Krishnan, B.: Isolated and dynamical horizons and their applications. Living Rev. Relat. **7**, 10 (2004)
7. Ashtekar, A., Beetle, C., Fairhurst, S.: Isolated horizons: a generalization of black hole mechanics. Class. Quantum Grav. **16**, L1 (1999)
8. Ashtekar, A., Beetle, C., Fairhurst, S.: Mechanics of isolated horizons. Class. Quantum Grav. **17**, 253 (2000)
9. Ashtekar, A., Beetle, C., Lewandowski, J.: Geometry of generic isolated horizons. Class. Quantum Gravity **19**, 1195 (2002)
10. Ashtekar, A., Beetle, C., Lewandowski, J.: Mechanics of rotating isolated horizons. Phys. Rev. D **64**, 044016 (2002)
11. Ashtekar, A., Corichi, A., Krasnov, K.: Isolated horizons: the classical phase space. Adv. Theor. Math. Phys. **3**, 419 (2000)
12. Ashtekar, A., Fairhurst, S., Krishnan, B.: Isolated horizons: Hamiltonian evolution and the first law. Phys. Rev. D **62**, 104025 (2000)
13. Ashtekar, A., Beetle, C., Dreyer, O., Fairhurst, S., Krishnan, B., Lewandowski, J., Wiśnieski, J.: Isolated horizons and their applications. Phys. Rev. Lett. **85**, 3564 (2000)
14. Bak, D., Rey, S.-J.: Cosmic holography. Class. Quantum Grav. **17**, L83 (2000)
15. Barceló, C., Liberati, S., Sonego, S., Visser, M.: Hawking-like radiation does not require a trapped region. Phys. Rev. Lett. **97**, 171301 (2006)
16. Baumgarte, T.W., Shapiro, S.L.: Numerical relativity and compact binaries. Phys. Rept. **376**, 41 (2003)
17. Bellucci, S., Faraoni, V.: Energy conditions and classical scalar fields. Nucl. Phys. B **640**, 453 (2002)
18. Bengtsson, I., Senovilla, J.M.M.: Region with trapped surfaces in spherical symmetry, its core, and their boundaries. Phys. Rev. D **83**, 044012 (2011)
19. Ben-Dov, I.: Outer trapped surfaces in Vaidya spacetimes. Phys. Rev. D **75**, 064007 (2007)
20. Bergmann, P.G.: Comments on the scalar tensor theory. Int. J. Theor. Phys. **1**, 25 (1968)
21. Booth, I.: Black hole boundaries. Can. J. Phys. **83**, 1073 (2005)
22. Booth, I., Fairhurst, S.: The first law for slowly evolving horizons. Phys. Rev. Lett. **92**, 011102 (2004)
23. Booth, I., Fairhurst, S.: Isolated, slowly evolving, and dynamical trapping horizons: Geometry and mechanics from surface deformations. Phys. Rev. D **75**, 084019 (2007)
24. Booth, I., Brits, L., Gonzalez, J.A., Van den Broeck, V.: Marginally trapped tubes and dynamical horizons. Class. Quantum Grav. **23**, 413 (2006)
25. Born, M.: Die theorie des starren elektrons in der kinematik des relativitätsprinzips. Ann. Physik (Leipzig) **335**, 1 (1909)
26. Brans, C., Dicke, R.H.: Mach's principle and a relativistic theory of gravitation. Phys. Rev. **124**, 925 (1961)
27. Carroll, S.M.: Spacetime and Geometry—An Introduction to General Relativity (Addison-Wesley, San Francisco, 2004)
28. Chriusciel, P.T.: Uniqueness of stationary, electro-vacuum black holes revisited. Helv. Physica Acta **69**, 529 (1996)
29. Chu, T., Pfeiffer, H.P., Cohen, M.I.: Horizon dynamics of distorted rotating black holes. Phys. Rev. D **83**, 104018 (2011)
30. Clifton, T.: Properties of black hole radiation from tunnelling. Class. Quantum Grav. **25**, 175022 (2008)
31. Collins, W.: Mechanics of apparent horizons. Phys. Rev. D **45**, 495 (1992)
32. Corichi, A., Sudarsky, D.: When is $S = A/4$? Mod. Phys. Lett. A **17**, 1431 (2002)

33. Csizmadia, P., Rácz, I.: Gravitational collapse and topology change in spherically symmetric dynamical systems. Class. Quantum Grav. **27**, 015001 (2010)
34. Davies, P.C.W.: Scalar production in Schwarzschild and Rindler metrics. J. Phys. A **8**, 609 (1975)
35. Di Criscienzo, R., Nadalini, M., Vanzo, L., Zerbini, S., Zoccatelli, G.: On the Hawking radiation as tunneling for a class of dynamical black holes. Phys. Lett. B **657**, 107 (2007)
36. Di Criscienzo, R., Hayward, S.A., Nadalini, M., Vanzo, L., Zerbini, S.: Hamilton-Jacobi method for dynamical horizons in different coordinate gauges. Class. Quantum Grav. **27**, 015006 (2010)
37. Dyer, C.C., Honig, E.: Conformal Killing horizons. J. Math. Phys. **20**, 409 (1979)
38. d'Inverno, R.: Introducing Einstein's Relativity. Oxford University Press, Oxford (2002)
39. Eardley, D.: Black hole boundary conditions and coordinate conditions. Phys. Rev. D **57**, 2299 (1998)
40. Eling, C., Guedens, R., Jacobson, T.: Nonequilibrium thermodynamics of spacetime. Phys. Rev. Lett. **96**, 121301 (2006)
41. Fairhurst, S., Krishnan, B.: Distorted black holes with charge. Int. J. Mod. Phys. D **10**, 691 (2001)
42. Figueras, P., Hubeny, V.E., Rangamani, M., Ross, S.F.: Dynamical black holes and expanding plasmas. J. High Energy Phys. **0904**, 137 (2009)
43. Fodor, G., Nakamura, K., Oshiro, Y., Tomimatsu, A.: Surface gravity in dynamical spherically symmetric space-times. Phys. Rev. D **54**, 3882 (1996)
44. Fulling, S.A.: Nonuniqueness of canonical field quantization in Riemannian space-time. Phys. Rev. D **7**, 2850 (1973)
45. Haijcek, P.: Origin of Hawking radiation. Phys. Rev. D **36**, 1065 (1987)
46. Hawking, S.W.: Gravitational radiation in an expanding universe. J. Math. Phys. **9**, 598 (1968)
47. Hawking, S.W.: Black holes in general relativity. Commun. Math. Phys. **25**, 152 (1972)
48. Hawking, S.W., G.Ellis, F.R.: The Large Scale Structure of Space-Time. Cambridge University Press, Cambridge (1973)
49. Hayward, S.A.: Quasilocal gravitational energy. Phys. Rev. D **49**, 831 (1994)
50. Hayward, S.A.: General laws of black hole dynamics. Phys. Rev. D **49**, 6467 (1994)
51. Hayward, S.A.: Gravitational energy in spherical symmetry. Phys. Rev. D **53**, 1938 (1996)
52. Hayward, S.A.: Unified first law of black-hole dynamics and relativistic thermodynamics. Class. Quantum Grav. **15**, 3147 (1998)
53. Hayward, S.A.: Formation and Evaporation of Nonsingular Black Holes. Phys. Rev. Lett. **96**, 031103 (2006)
54. Hayward, S.A., Di Criscienzo, R., Nadalini, M., Vanzo, L., Zerbini, S.: Local Hawking temperature for dynamical black holes. Class. Quantum Grav. **26**, 062001 (2009)
55. Hernandez, W.C., Misner, C.W.: Observer time as a coordinate in relativistic spherical hydrodynamics. Astrophys. J. **143**, 452 (1966)
56. Hiscock, W.A.: Gravitational entropy of nonstationary black holes and spherical shells. Phys. Rev. D **40**, 1336 (1989)
57. Jacobson, T.: Thermodynamics of spacetime: the Einstein equation of state. Phys. Rev. Lett. **75**, 1260 (1995)
58. Jang, K.-X., Feng, T., Peng, D.-T.: Hawking radiation of apparent horizon in a FRW universe as tunneling beyond semiclassical approximation. Int. J. Theor. Phys. **48**, 2112 (2009)
59. Kavanagh, W., Booth, I.: Spacetimes containing slowly evolving horizons. Phys. Rev. D **74**, 044027 (2006)
60. Kodama, H.: Conserved energy flux from the spherically symmetric system and the back reaction problem in the black hole evaporation. Progr. Theor. Phys. **63**, 1217 (1980)
61. McClure, M.L., Dyer, C.C.: Asymptotically Einstein-de Sitter cosmological black holes and the problem of energy conditions. Class. Quantum Grav. **23**, 1971 (2006)
62. McClure, M.L., Anderson, K., Bardahl, K.: Cosmological versions of Vaidya's radiating stellar exterior, an accelerating reference frame, and Kinnersley's photon rocket. Preprint arXiv:0709.3288

63. McClure, M.L., Anderson, K., Bardahl, K.: Nonisolated dynamical black holes and white holes. Phys. Rev. D **77**, 104008 (2008)
64. Misner, C.W., Sharp, D.H.: Relativistic equations for adiabatic, spherically symmetric gravitational collapse. Phys. Rev. **136**, 571 (1964)
65. Morris, M.S., Thorne, K.S.: Wormholes in space-time and their use for interstellar travel: a tool for teaching general relativity. Am. J. Phys. **56**, 395 (1988)
66. Mukohyama, S., Hayward, S.A.: Quasilocal first law of black hole dynamics. Class. Quantum Grav. **17**, 2153 (2000)
67. Nielsen, A.B.: Black holes without boundaries. Int. J. Mod. Phys. D **17**, 2359 (2009)
68. Nielsen, A.B.: Black holes and black hole thermodynamics without event horizons. Gen. Rel. Gravit. **41**, 1539 (2009)
69. Nielsen, A.B.: The spatial relation between the event horizon and trapping horizon. Class. Quantum Gravity **27**, 245016 (2010)
70. Nielsen, A.B., Firouzjaee, J.T.: Conformally rescaled spacetimes and Hawking radiation. Gen. Rel. Gravit. **45**, 1815 (2013)
71. Nielsen, A.B., Yeom, D.-H.: Spherically symmetric trapping horizons, the Misner-Sharp mass and black hole evaporation. Int. J. Mod. Phys. A **24**, 5261 (2009)
72. Nielsen, A.B., Yoon, J.H.: Dynamical surface gravity. Class. Quantum Grav. **25**, 085010 (2008)
73. Nielsen, A.B., Visser, M.: Production and decay of evolving horizons. Class. Quantum Grav. **23**, 4637 (2006)
74. Nordtvedt, K.: PostNewtonian metric for a general class of scalar tensor gravitational theories and observational consequences. Astrophys. J. **161**, 1059 (1970)
75. Parikh, M.K., Wilczek, F.: Hawking radiation as tunneling. Phys. Rev. Lett. **85**, 5042 (2000)
76. Penrose, R.: Gravitational collapse and spacetime singularities. Phys. Rev. Lett. **14**, 57 (1965)
77. Pielahn, M., Kunstatter, G., Nielsen, A.B.: Critical analysis of dynamical surface gravity in spherically symmetric black hole formation. Phys. Rev. D **84**, 104008 (2011)
78. Poisson, E.: A Relativist's Toolkit: The Mathematics of Black-Hole Mechanics. Cambridge University Press, Cambridge (2004)
79. Rácz, I.: On the use of the Kodama vector field in spherically symmetric dynamical problems. Class. Quantum Grav. **23**, 115 (2006)
80. Roman, T.A., Bergmann, P.G.: Stellar collapse without singularities? Phys. Rev. D **28**, 1265 (1983)
81. Scheel, M.A., Shapiro, S.L., Teukolsky, S.A.: Collapse to black holes in Brans-Dicke theory. 2. Comparison with general relativity. Phys. Rev. D **51**, 4236 (1995)
82. Schnetter, E., Krishnan, B.: Non-symmetric trapped surfaces in the Schwarzschild and Vaidya spacetimes. Phys. Rev. D **73**, 021502 (2006)
83. Sorkin, R.D.: In: Wiltshire, D. (ed.) Proceedings of the First Australasian Conference on General Relativity and Gravitation, February 1996, Adelaide, pp. 163–174. University of Adelaide (1996). [Preprint arXiv:gr-qc/9701056]
84. Sultana, J., Dyer, C.C.: Conformal killing horizons. J. Math. Phys. **45**, 4764 (2004)
85. Sultana, J., Dyer, C.C.: Cosmological black holes: A black hole in the Einstein-de Sitter universe. Gen. Rel. Gravit. **37**, 1349 (2005)
86. Szabados, L.: Quasi-local energy-momentum and angular momentum in GR: a review article. Living Rev. Relat. **7**, 4 (2004)
87. Tolman, R.C.: Non-Newtonian Mechanics. Some transformation equations. Philos. Mag. **25** (125), 150 (1912)
88. Tung, R.-S.: Stationary untrapped boundary conditions in general relativity. Class. Quantum Grav. **25**, 085005 (2008)
89. Unruh, W.G.: Notes on black-hole evaporation. Phys. Rev. D **14**, 870 (1976)
90. Vanzo, L., Acquaviva, G., Di Criscienzo, R.: Tunnelling methods and Hawking's radiation: achievements and prospects. Class. Quantum Grav. **28**, 183001 (2011)
91. Visser, M.: Dirty black holes: thermodynamics and horizon structure. Phys. Rev. D **46**, 2445 (1992)

92. Visser, M.: Gravitational vacuum polarization. I. Energy conditions in the Hartle-Hawking vacuum. Phys. Rev. D **54**, 5103 (1996)
93. Visser, M.: Essential and inessential features of Hawking radiation. Int. J. Mod. Phys. D **12**, 649 (2003)
94. Wagoner, R.V.: Scalar-tensor theory and gravitational waves. Phys. Rev. D **1**, 3209 (1970)
95. Wald, R.M.: General Relativity. Chicago University Press, Chicago (1984)
96. Wald, R.M.: The thermodynamics of black holes. Living Rev. Relat. **4**, 6 (2001)
97. Wald, R.M., Iyer, V.: Trapped surfaces in the Schwarzschild geometry and cosmic censorship. Phys. Rev. D **44**, R3719 (1991)

Chapter 3
Cosmological Horizons

Simplicity is the ultimate sophistication.

—Leonardo da Vinci

3.1 Introduction

Friedmann-Lemaître-Robertson-Walker (FLRW) spacetimes are spherically symmetric about every spatial point and are, therefore, trivially spherically symmetric, but are nevertheless important. These cosmological spacetimes are much simpler than black hole spacetimes but still contain horizons. Cosmological horizons have been studied in inflationary scenarios of the early universe in relation to the so-called horizon problem. In general, FLRW spaces contain time-dependent apparent horizons expressed by particularly simple equations and are interesting from our point of view. Due to their simplicity, it is convenient to discuss these apparent horizons before approaching the more complicated horizons of time-dependent black holes in the following chapters.

We begin by using an analogy between black hole and cosmological horizons and examining various coordinate systems for FLRW spaces which mimic the corresponding ones for Schwarzschild black holes. Several coordinate systems are employed in the study of Hawking radiation from black hole horizons and different coordinates, or various families of observers, provide potentially different notions of surface gravity and horizon temperature. The underlying idea is to develop coordinate systems useful to study Hawking radiation from cosmological horizons and, later, from the time-dependent horizons of black holes embedded in time-varying cosmological backgrounds. We also compute useful geometric quantities in these various coordinate systems.

© Springer International Publishing Switzerland 2015
V. Faraoni, *Cosmological and Black Hole Apparent Horizons*,
Lecture Notes in Physics 907, DOI 10.1007/978-3-319-19240-6_3

3.1.1   FLRW Cosmologies

A FLRW space has line element

$$ds^2 = -dt^2 + a^2(t) \left(\frac{dr^2}{1 - kr^2} + r^2 d\Omega_{(2)}^2 \right) \tag{3.1}$$

in comoving (or "synchronous") coordinates (t, r, θ, φ), where k is the curvature index and $a(t)$ is the scale factor. The comoving coordinate r is not an areal radius; the latter is instead a function of both t and r,

$$R(t, r) = a(t)r. \tag{3.2}$$

It is also common to use the conformal time η defined by $dt = a d\eta$, in terms of which

$$ds^2 = a^2(\eta) \left(-d\eta^2 + \frac{dr^2}{1 - kr^2} + r^2 d\Omega_{(2)}^2 \right). \tag{3.3}$$

In the spatially flat case $k = 0$ this line element is explicitly conformal to the Minkowski one, however *all* FLRW spaces are conformally flat since their Weyl tensor vanishes identically. The conformal flatness of the FLRW line element for $k \neq 0$ can be made explicit by transforming to suitable coordinates ([31, 48, 49] and references therein).

In General Relativity, the Einstein equations reduce to ordinary differential equations in a FLRW space. If the FLRW universe is sourced by a perfect fluid with energy-momentum tensor

$$T_{ab} = (P + \rho) u_a u_b + P g_{ab}, \tag{3.4}$$

where ρ, P, and u^a are the energy density, pressure, and 4-velocity field of the fluid, respectively, one has

$$H^2 = \frac{8\pi}{3} \rho - \frac{k}{a^2}, \tag{3.5}$$

$$\frac{\ddot{a}}{a} = \dot{H} + H^2 = -\frac{4\pi}{3} (\rho + 3P) \tag{3.6}$$

(Einstein-Friedmann equations), where an overdot denotes differentiation with respect to the comoving time t and $H(t) \equiv \dot{a}/a$ is the Hubble parameter. The covariant conservation equation $\nabla^b T_{ab} = 0$ then yields the energy conservation equation

$$\dot{\rho} + 3H (P + \rho) = 0. \tag{3.7}$$

This equation is not independent of Eqs. (3.5) and (3.6) (in the same way that the covariant conservation equation $\nabla^b T_{ab} = 0$ is not independent of the Einstein equations) and can be derived from them. Another useful relation which follows from Eqs. (3.5) and (3.6) is

$$\dot{H} = -4\pi \, (P + \rho) + \frac{k}{a^2} \, . \tag{3.8}$$

The expression of the Ricci curvature in terms of the Hubble parameter and its derivative is useful:

$$\mathscr{R} = 6\left(\dot{H} + 2H^2 + \frac{k}{a^2}\right) . \tag{3.9}$$

Let $t = 0$ denote the Big Bang spacetime singularity (in the cases in which it is present). All comoving observers whose worldlines have u^a as tangent are physically equivalent and, therefore, the following considerations apply to any of them, although we refer explicitly to a comoving observer located at $r = 0$.

3.2 Hyperspherical Coordinates for FLRW Space

The FLRW line element can be written using hyperspherical coordinates

$$ds^2 = -dt^2 + a^2(t)\left[d\chi^2 + f^2(\chi)d\Omega^2_{(2)}\right], \tag{3.10}$$

where

$$f(\chi) = r = \begin{cases} \sinh \chi & \text{if } k < 0, \\ \chi & \text{if } k = 0, \\ \sin \chi & \text{if } k > 0, \end{cases} \tag{3.11}$$

and

$$\chi = f^{-1}(r) = \int \frac{dr}{\sqrt{1 - kr^2}} \, . \tag{3.12}$$

3.3 Kruskal-Szekeres Coordinates for de Sitter Space

Kruskal-Szekeres coordinates for de Sitter space were introduced by Gibbons and Hawking [41] in their study of the thermodynamics of the de Sitter event horizon. These (U, V) coordinates are defined by [41]

$$Hr = 1 - U^2V^2, \tag{3.13}$$

$$2Ht = \ln\left(-\frac{U}{V}\right) . \tag{3.14}$$

Kruskal-Szekeres coordinates for FLRW spaces other than de Sitter are not known.

3.4   Painlevé-Gullstrand and Schwarzschild-Like Coordinates for $k = 0$ FLRW Space

Consider the spatially flat $(k = 0)$ FLRW metric in comoving coordinates

$$ds^2 = -dt^2 + a^2(t) \left(dr^2 + r^2 d\Omega_{(2)}^2 \right) ;$$

upon use of the areal radius $R \equiv a(t)r$ and of the relation between differentials

$$dr = \frac{dR}{a} - Hr\, dt = \frac{1}{a} (dR - HRdt) , \tag{3.15}$$

the metric is recast in the *Painlevé-Gullstrand form*[1]

$$ds^2 = - \left(1 - H^2R^2\right) dt^2 - 2HR\, dtdR + dR^2 + R^2 d\Omega_{(2)}^2 . \tag{3.16}$$

This form is useful for comparison with solutions of the field equations describing black holes or central objects embedded in a spatially flat FLRW background. The history and the advantages of Painlevé-Gullstrand coordinates in the study of black hole horizons are discussed in Refs. [55, 59, 68] (see [32] for Painlevé-Gullstrand coordinates in Kerr spacetime). We can eliminate the cross-term proportional to $dtdR$ by introducing the new time coordinate T defined by

$$dT = \frac{1}{F} (dt + \beta dR) , \tag{3.17}$$

where $F(t, R)$ is an integrating factor satisfying[2]

$$\frac{\partial}{\partial R} \left(\frac{1}{F} \right) = \frac{\partial}{\partial t} \left(\frac{\beta}{F} \right) \tag{3.18}$$

to guarantee that dT is a locally exact differential, and $\beta(t, r)$ is a function to be determined. Then, $dt = FdT - \beta dR$ and, substituting into the line element (3.16), one obtains

$$\begin{aligned} ds^2 &= - \left(1 - H^2R^2\right) \left(F^2 dT^2 + \beta^2 dR^2 - 2F\beta dTdR \right) \\ &\quad - 2HRdR \left(FdT - \beta dR \right) + dR^2 + R^2 d\Omega_{(2)}^2 \\ &= - \left(1 - H^2R^2\right) F^2 dT^2 + 2F \left[\left(1 - H^2R^2\right) \beta - HR \right] dTdR \\ &\quad + \left[1 - \left(1 - H^2R^2\right) \beta^2 + 2\beta HR \right] dR^2 + R^2 d\Omega_{(2)}^2 . \end{aligned} \tag{3.19}$$

---

[1] Also called "r-gauge" (e.g., [30]).

[2] As in most situations, the integrating factor is not unique.

By setting

$$\beta(t, R) = \frac{HR}{1 - H^2 R^2} \tag{3.20}$$

one obtains the FLRW metric in the *Schwarzschild-like form*[3]

$$ds^2 = -\left(1 - H^2 R^2\right) F^2 dT^2 + \frac{dR^2}{1 - H^2 R^2} + R^2 d\Omega_{(2)}^2, \tag{3.21}$$

which is again reminiscent of the Schwarzschild line element, except for the presence of the factor F and for the fact that the Hubble parameter H is not constant (unless the FLRW space reduces to de Sitter space, in which case $F \equiv 1$). This line element is of the form (2.94). By comparing Eqs. (3.21) and (2.94), or using directly the definition (2.92), one obtains the Misner-Sharp-Hernandez mass

$$M_{\mathrm{MSH}} = \frac{H^2 R^3}{2} \doteq \frac{4\pi \rho R^3}{3} \tag{3.22}$$

(which matches Eq. (3.56) below and constitutes a consistency check) and

$$e^{-\phi} = F. \tag{3.23}$$

Here an overdot on the equality sign denotes the fact that this equality holds in General Relativity in a FLRW universe sourced by a perfect fluid. The cosmological apparent horizon of a spatially flat FLRW space is now easily located by setting $g^{RR} = 0$, which yields

$$R_{\mathrm{AH}} = \frac{1}{H}. \tag{3.24}$$

3.5 Schwarzschild-Like Coordinates for General FLRW Spaces

Painlevé-Gullstrand coordinates are useful in the discussion of black hole horizons because they are regular across these horizons, while Schwarzschild-like coordinates (the ones commonly used to introduce the Schwarzschild solution and black

---

[3]The coordinate system in which the metric assumes the form (3.21) is sometimes referred to as *Nolan gauge* in the literature. Nolan [69–71] studied this coordinate system in the case of the McVittie metric [61] representing a central object embedded in a FLRW universe, eventually restricting to $k = 0$ or -1. The McVittie metric reduces to that of Eq. (3.21) in the limit in which the mass of the central object vanishes, hence (3.21) is a trivial case of the McVittie line element in the Nolan gauge.

holes) are not. A similar situation occurs for cosmological horizons, although Schwarzschild-like coordinates are not so obvious in this case (in fact, most people are unfamiliar with them, except for the case of de Sitter and Schwarzschild-de Sitter spaces) and usually cosmology is formulated using comoving or hyperspherical coordinates. In the following we derive pseudo-Painlevé-Gullstrand and Schwarzschild-like coordinates for a general FLRW space.

Begin from the FLRW metric in comoving coordinates (3.1); using the areal radius $R(t, r) \equiv a(t)r$ and Eq. (3.15), the FLRW line element (3.1) assumes the *pseudo-Painlevé-Gullstrand form*

$$ds^2 = -\left(1 - \frac{H^2 R^2}{1 - kR^2/a^2}\right) dt^2 - \frac{2HR}{1 - kR^2/a^2} \, dt dR + \frac{dR^2}{1 - kR^2/a^2} + R^2 d\Omega_{(2)}^2 .$$

$$(3.25)$$

In the absence of a better nomenclature, we use the name "pseudo Painlevé-Gullstrand coordinates" because the coefficient of dR^2 is not unity, as required for Painlevé-Gullstrand coordinates[4] and the spacelike surfaces $t = $ constant are not flat (unless $k = 0$), which is instead the essential property of Painlevé-Gullstrand coordinates [59]. By using the fact that

$$1 - \frac{H^2 R^2}{1 - kr^2} = \frac{1 - R^2/R_{AH}^2}{1 - kr^2}$$

with

$$R_{AH} \equiv \frac{1}{\sqrt{H^2 + k/a^2}},$$

$$(3.26)$$

one can write the line element (3.25) as [20, 51, 56]

$$ds^2 = -\frac{1 - R^2/R_{AH}^2}{1 - kR^2/a^2} \, dt^2 - \frac{2HR}{1 - kR^2/a^2} \, dt dR + \frac{dR^2}{1 - kR^2/a^2} + R^2 d\Omega_{(2)}^2 . \quad (3.27)$$

Setting $ds^2 = 0$ and $d\theta = d\varphi = 0$ for radial null rays with tangents p^a yields

$$p^1 = HR \pm \sqrt{H^2 R^2 + 1 - \frac{R^2}{R_{AH}^2}}$$

$$(3.28)$$

with the choice $p^0 = 1$ (e.g., [20]).

In Painlevé-Gullstrand coordinates, a common trick consisting of setting $p^1 = dR/d\lambda = 0$ to locate the apparent horizon is based on a rigorous argument and does indeed provide the correct result (this argument holds true for general spherically

---

[4]The literature contains ambiguous terminology for general FLRW spaces (e.g., [20]), while the de Sitter case does not lend itself to these ambiguities [62, 72].

symmetric metrics in Painlevé-Gullstrand coordinates [68]). A posteriori, this trick works also for quasi-Painlevé-Gullstrand coordinates in our specific situation.

To transform to the Schwarzschild-like form, we first introduce the new time coordinate T defined by

$$dT = \frac{1}{F} (dt + \beta dR) , \qquad (3.29)$$

where F is a (non-unique) integrating factor satisfying again

$$\frac{\partial}{\partial R} \left(\frac{1}{F} \right) = \frac{\partial}{\partial t} \left(\frac{\beta}{F} \right) \qquad (3.30)$$

to guarantee that dT is a locally exact differential, while $\beta(t, R)$ is a function to be determined. Substituting $dt = FdT - \beta dR$ into the line element, one obtains

$$ds^2 = - \left(1 - \frac{H^2 R^2}{1 - kr^2} \right) F^2 dT^2$$

$$+ \left[- \left(1 - \frac{H^2 R^2}{1 - kr^2} \right) \beta^2 + \frac{2HR\beta}{1 - kr^2} + \frac{1}{1 - kr^2} \right] dR^2$$

$$+ 2 \left(1 - \frac{H^2 R^2}{1 - kr^2} \right) F\beta dTdR - \frac{2HRF}{1 - kr^2} dTdR + R^2 d\Omega_{(2)}^2 . \qquad (3.31)$$

By choosing

$$\beta = \frac{HR}{\left(1 - \frac{H^2 R^2}{1 - kr^2} \right) (1 - kr^2)} = \frac{HR}{1 - H^2 R^2 - kr^2} , \qquad (3.32)$$

the cross-term proportional to $dTdR$ is eliminated and we obtain the *FLRW line element in the Schwarzschild-like form*

$$ds^2 = - \left(1 - \frac{H^2 R^2}{1 - kR^2/a^2} \right) F^2 dT^2 + \frac{dR^2}{1 - kR^2/a^2 - H^2 R^2} + R^2 d\Omega_{(2)}^2 , \qquad (3.33)$$

where $F(T, R)$, a, and H are implicit functions of T. By using the expression of the Misner-Sharp-Hernandez mass in FLRW space (3.56) below, this line element can be cast as

$$ds^2 = - \left(\frac{1 - \frac{2M}{R}}{1 - \frac{2M}{R} + H^2 R^2} \right) F^2 dT^2 + \frac{dR^2}{1 - \frac{2M}{R}} + R^2 d\Omega_{(2)}^2 . \qquad (3.34)$$

A spherically symmetric metric can always be put in the form (2.94) [68]. By comparing Eqs. (3.33) and (2.94), it follows that

$$e^{-\phi} = \frac{F(T,R)}{\sqrt{1 - kR^2/a^2}}$$

and

$$1 - \frac{2M_{\mathrm{MSH}}}{R} = 1 - \frac{kR^2}{a^2} - H^2 R^2 \tag{3.35}$$

(this equation is consistent with the expression (3.56) of the Misner-Sharp-Hernandez mass in FLRW space that we are going to discuss soon).

It is now easy to locate the apparent horizon of a general FLRW space using the prescription (2.104). Setting $g^{RR}|_{\mathrm{AH}} = 0$ and reading $g^{RR} = g_{RR}^{-1}$ from Eq. (3.33) yield the radius of the FLRW apparent horizon

$$R_{\mathrm{AH}} = \frac{1}{\sqrt{H^2 + k/a^2}}. \tag{3.36}$$

The components of the Kodama vector in Schwarzschild-like coordinates are given by Eq. (2.76), which yields

$$K^{\mu} = \left(-\frac{\sqrt{1 - kR^2/a^2}}{F}, 0, 0, 0 \right) \tag{3.37}$$

and its norm squared is

$$K_c K^c = -\left(1 - H^2 R^2 - \frac{kR^2}{a^2} \right) = -\left(1 - \frac{R^2}{R_{\mathrm{AH}}^2} \right). \tag{3.38}$$

The Kodama vector is timelike ($K_c K^c < 0$) if $R < R_{\mathrm{AH}}$, null if $R = R_{\mathrm{AH}}$, and spacelike ($K_c K^c > 0$) outside the apparent horizon $R > R_{\mathrm{AH}}$. The Kodama vector produces the Misner-Sharp-Hernandez mass as a Noether charge [45]. Equations (3.35) and the Hamiltonian constraint (3.5) then imply that

$$M_{\mathrm{MSH}}(R) = \frac{4\pi R^3}{3} \rho. \tag{3.39}$$

It is often useful to know the components of the Kodama vector in pseudo-Painlevé-Gullstrand coordinates recurrent in the literature (e.g., [20]) and in comoving coordinates, which we report here (see Appendix A.2 for details):

$$K^{\mu} = \left(-\sqrt{1 - kR^2/a^2}, 0, 0, 0 \right) \tag{3.40}$$

in pseudo-Painlevé-Gullstrand coordinates, and

$$K^\mu = \left(-\sqrt{1 - kr^2}, Hr\sqrt{1 - kr^2}, 0, 0\right) \tag{3.41}$$

in comoving coordinates (t, r, θ, φ). The norm squared of the Kodama vector is

$$K^c K_c = -\left(1 - kr^2 - \dot{a}^2 r^2\right) = -\left(1 - 2M/R\right) . \tag{3.42}$$

3.6 Painlevé-Gullstrand Coordinates for General FLRW Spaces

To find Painlevé-Gullstrand coordinates for general (i.e., not necessarily spatially flat) FLRW spacetimes, begin from the FLRW line element in Schwarzschild-like coordinates (3.33). We search for a new time coordinate $\tilde{T}(T, R)$, with

$$d\tilde{T} = \frac{\partial \tilde{T}}{\partial T} dT + \frac{\partial \tilde{T}}{\partial R} dR \equiv \dot{\tilde{T}} dT + \tilde{T}' dR . \tag{3.43}$$

By substituting

$$dT = \frac{1}{\dot{\tilde{T}}} d\tilde{T} - \frac{\tilde{T}'}{\dot{\tilde{T}}} dR \tag{3.44}$$

one obtains

$$ds^2 = -\left(1 - \frac{H^2 R^2}{1 - kR^2/a^2}\right) F^2 \left(\frac{d\tilde{T}}{\dot{\tilde{T}}} - \frac{\tilde{T}'}{\dot{\tilde{T}}} dR\right)^2 + \frac{dR^2}{1 - H^2R^2 - kR^2/a^2}$$

$$+ R^2 d\Omega_{(2)}^2$$

$$= -\left(1 - \frac{H^2 R^2}{1 - kR^2/a^2}\right) \left(\frac{F}{\dot{\tilde{T}}}\right)^2 d\tilde{T}^2$$

$$+ \left[-\left(1 - \frac{H^2 R^2}{1 - kR^2/a^2}\right) \left(\frac{F\tilde{T}'}{\dot{\tilde{T}}}\right)^2 + \frac{1}{1 - H^2R^2 - kR^2/a^2}\right] dR^2$$

$$+ \frac{2F^2 \tilde{T}'}{\dot{\tilde{T}}^2} \left(1 - \frac{H^2 R^2}{1 - kR^2/a^2}\right) d\tilde{T} dR + R^2 d\Omega_{(2)}^2 . \tag{3.45}$$

By imposing that the coefficient of dR^2 be unity one obtains

$$\tilde{T}' = \pm \frac{\dot{\tilde{T}} R \sqrt{\left(H^2 + \frac{k}{a^2}\right) \left(1 - \frac{kR^2}{a^2}\right)}}{F \left(1 - H^2 R^2 - \frac{kR^2}{a^2}\right)} \tag{3.46}$$

which, substituted into the previous equation, yields the *FLRW line element in Painlevé-Gullstrand coordinates*

$$ds^2 = -\left(1 - \frac{H^2R^2}{1 - kR^2/a^2}\right)\left(\frac{F}{\dot{\tilde{T}}}\right)^2 d\tilde{T}^2 \pm \frac{2FR}{\dot{\tilde{T}}}\sqrt{\frac{H^2 + k/a^2}{1 - kR^2/a^2}}\, d\tilde{T}dR + dR^2$$

$$+R^2 d\Omega_{(2)}^2\,. \tag{3.47}$$

Clearly, slicings of constant time $\tilde{T}$ are flat.

3.7   Congruences of Radial Null Geodesics in FLRW Space

Having given the general definitions and having calculated the needed formulae, let us move to the study of congruences of radial null geodesics and of cosmological horizons. In FLRW space, which is spherically symmetric about every point of space, the outgoing and ingoing radial null geodesics have tangent fields with comoving components

$$l^\mu = \left(1, \frac{\sqrt{1 - kr^2}}{a(t)}, 0, 0\right)\,, \qquad n^\mu = \left(1, -\frac{\sqrt{1 - kr^2}}{a(t)}, 0, 0\right)\,, \tag{3.48}$$

respectively.

Proof. Setting $p_c p^c = 0$ for the 4-tangents p^a yields

$$0 = p_c p^c = -(p^0)^2 + \frac{a^2}{1 - kr^2}\,(p^1)^2$$

and

$$p^1 = \pm\frac{\sqrt{1 - kr^2}}{a}\,p^0\,.$$

Of course, one can also set $ds^2 = 0$ and $d\theta = d\varphi = 0$, obtaining directly

$$\frac{dr}{dt} = \frac{dr/d\lambda}{dt/d\lambda} = \frac{p^1}{p^0} = \pm\frac{\sqrt{1 - kr^2}}{a}\,, \tag{3.49}$$

where λ is a parameter along the null rays. $\square$

There is freedom to rescale a future-directed null vector by an arbitrary regular function (which must be positive if we want to keep this vector future-oriented). The choice of normalization in Eq. (3.48) implies that $l^c n_c = -2$. The more common normalization $l^c n_c = -1$ is obtained by dividing both l^a and n^a by $\sqrt{2}$.

The expansions of these null geodesic congruences are computed using Eq. (2.27). One first computes

$$\nabla_c l^c = \frac{1}{\sqrt{-g}} \partial_\mu \left(\sqrt{-g}\, l^\mu \right) = \frac{\left(1 - kr^2\right)^{1/2}}{a^3 r^2} \left[\frac{3a^2 \dot{a} r^2}{\left(1 - kr^2\right)^{1/2}} + 2a^2 r \right]$$

$$= 3H + \frac{2}{ar} \sqrt{1 - kr^2}\,, \tag{3.50}$$

$$\nabla_c n^c = \frac{1}{\sqrt{-g}} \partial_\mu \left(\sqrt{-g}\, n^\mu \right) = \frac{\left(1 - kr^2\right)^{1/2}}{a^3 r^2} \left[\frac{3a^2 \dot{a} r^2}{\left(1 - kr^2\right)^{1/2}} - 2a^2 r \right]$$

$$= 3H - \frac{2}{ar} \sqrt{1 - kr^2}\,. \tag{3.51}$$

Then, using

$$\sqrt{-g} = \frac{a^3 r^2 \sin^2 \theta}{\sqrt{1 - kr^2}}\,,$$

$$\Gamma^c_{00} = 0\,, \quad \Gamma^c_{01} = \Gamma^c_{10} = H\delta^{c1}\,, \quad \Gamma^c_{11} = \frac{kr\delta^{c1} + a\dot{a}\,\delta^{c0}}{1 - kr^2}\,,$$

$$g_{cd}\, l^c n^d = -2\,,$$

Eq. (2.27) yields[5]

$$\theta_l = \frac{2\left(\dot{a} r + \sqrt{1 - kr^2} \right)}{ar} = 2 \left(H + \frac{1}{R} \sqrt{1 - \frac{kR^2}{a^2}} \right)\,, \tag{3.52}$$

$$\theta_n = \frac{2\left(\dot{a} r - \sqrt{1 - kr^2} \right)}{ar} = 2 \left(H - \frac{1}{R} \sqrt{1 - \frac{kR^2}{a^2}} \right)\,. \tag{3.53}$$

The cosmological apparent horizon is located where $\theta_n = 0$ and $\theta_l > 0$, i.e., at

$$r_{\mathrm{AH}} = \frac{1}{\sqrt{\dot{a}^2 + k}}\,, \tag{3.54}$$

or

$$R_{\mathrm{AH}}(t) = \frac{1}{\sqrt{H^2 + k/a^2}} \tag{3.55}$$

---

[5]See, e.g., Ref. [36]. The factor 2 in Eqs. (3.52) and (3.53) does not appear in Ref. [5] and in other works because of different normalizations of l^a and n^a. The expansions of the congruences definitely depend on the choice made for $l^c n_c$.

in terms of the proper radius $R \equiv ar$. Note that the apparent horizon is defined using only null geodesic congruences and their expansions and there is no reference to the global causal structure.

Sometimes it is tempting to locate apparent horizons by simply guessing "where the outgoing radial null rays stop", that is, by setting $l^r = 0$. Although this shortcut may sometimes provide the correct result (it does work for spherically symmetric metrics in Painlevé-Gullstrand coordinates [68]), in general, it is not to be adopted in place of the proper procedure which consists of finding the surfaces on which $\theta_l = 0$ and $\theta_n \neq 0$. The radial null geodesic congruences in FLRW space offer a counterexample: using Eq. (3.48) and setting $n^1 = 0$ would lead to the incorrect conclusion that there are no apparent or trapping horizons in $k = 0, -1$ FLRW spaces, and to the incorrect value of R_{AH} for $k = +1$ FLRW space. This is obviously incorrect: except for the Minkowski case $H = 0$, apparent horizons always exist and are given by Eq. (3.55).

3.8   Horizons in FLRW Space

Two horizons of FLRW space are familiar from standard cosmology textbooks: they are the particle horizon and the event horizon [75]. In addition, apparent and trapping horizons are relevant for our discussion. Consider a FLRW universe with line element (3.1) in comoving coordinates. The proper, or areal, radius is $R \equiv a(t)r$ and the Misner-Sharp-Hernandez mass of a sphere of radius R, defined by Eq. (2.92), is easily found to be

$$M_{\mathrm{MSH}} = \left(H^2 + \frac{k}{a^2} \right) \frac{R^3}{2} \doteq \frac{4\pi R^3}{3}\, \rho\,, \tag{3.56}$$

where, again, a dot on the last equality denotes the fact that it holds in General Relativity in a universe sourced by a perfect fluid (note that Eq. (3.56) is valid for any value of the curvature index k). In non-spatially flat FLRW spaces, $k \neq 0$, the quantity $4\pi R^3/3$ is *not* the proper volume of a sphere of radius R, which is instead

$$V_{\mathrm{proper}} = \int_0^{2\pi} d\varphi \int_0^\pi d\theta \int_0^r dr' \sqrt{g^{(3)}}\,, \tag{3.57}$$

where $g^{(3)} = \dfrac{a^6 r^4 \sin^2 \theta}{1 - kr^2}$ is the determinant of the restriction of the metric g_{ab} to the 3-surfaces $t = $ constant. Therefore,

$$V_{\mathrm{proper}} = 4\pi a^3(t) \int_0^r \frac{dr' r'^2}{\sqrt{1 - kr'^2}} = 4\pi a^3(t) \int_0^\chi d\chi' f^2(\chi')\,, \tag{3.58}$$

where χ is the hyperspherical coordinate and the function $f(\chi)$ is given by Eq. (3.11). However, it turns out that only the "areal volume"

$$V \equiv \frac{4\pi R^3}{3} \tag{3.59}$$

will be needed for our purposes, as a consequence of the use of the Misner-Sharp-Hernandez mass which is usually identified with the internal energy U that appears in the thermodynamics of the apparent horizon.

Proof. The FLRW line element can be rewritten in Schwarzschild-like coordinates as (Sect. 3.5)

$$ds^2 = -\left(1 - \frac{H^2 R^2}{1 - kR^2/a^2}\right) F^2 dt^2 + \frac{dR^2}{1 - kR^2/a^2 - H^2 R^2} + R^2 d\Omega_{(2)}^2,$$

where $F(T, R)$ is an integrating factor. Equation (2.92) then gives

$$1 - \frac{2M_{\mathrm{MSH}}}{R} = g^{RR} = 1 - H^2 R^2 - \frac{kR^2}{a^2}$$

and

$$M_{\mathrm{MSH}} = \left(H^2 + \frac{k}{a^2}\right) \frac{R^3}{2} \doteq \frac{4\pi R^3}{3} \rho, \tag{3.60}$$

using the Hamiltonian constraint (3.5) valid in General Relativity with a perfect fluid.   □

The Misner-Sharp-Hernandez mass of a sphere of radius R does not depend explicitly on the pressure P of the cosmic fluid. Its time derivative, instead, does: consider a sphere of proper radius $R = R_{\mathrm{s}}(t)$, then, using $R \equiv a(t)r$ and Eq. (3.7), one has

$$\dot{M}_{\mathrm{MSH}} \equiv \frac{dM_{\mathrm{MSH}}}{dt} = \frac{4\pi}{3}\left(3R_{\mathrm{s}}^2 \dot{R}_{\mathrm{s}}\rho + R_{\mathrm{s}}^3 \dot{\rho}\right)$$

$$= \frac{4\pi}{3}\left\{3R_{\mathrm{s}}^2 \dot{R}_{\mathrm{s}}\rho + R_{\mathrm{s}}^3\left[-3H(P + \rho)\right]\right\}$$

$$= 4\pi\left[R_{\mathrm{s}}^2 \dot{R}_{\mathrm{s}}\rho - HR_{\mathrm{s}}^3(P + \rho)\right]$$

$$= 4\pi R_{\mathrm{s}}^3\left[\frac{\dot{R}_{\mathrm{s}}}{R_{\mathrm{s}}}\rho - H(P + \rho)\right].$$

If the sphere is comoving with radius $R_{\mathrm{s}} \propto a(t)$ then $\dot{R}_{\mathrm{s}}/R_{\mathrm{s}} = H$ and

$$\dot{M}_{\mathrm{MSH}} = -4\pi H R_{\mathrm{s}}^3 P; \tag{3.61}$$

in this case $\dot{M}_{MSH}$ depends explicitly on P but not on ρ. By taking the ratio of Eqs. (3.61) and (3.56) one also obtains that, in General Relativity,

$$\dot{M}_{MSH} + 3H \frac{P}{\rho} M_{MSH} = 0 \qquad \text{(comoving sphere)}. \tag{3.62}$$

3.8.1   Particle Horizon

The *particle horizon* [75] at time t is a sphere centered on the comoving observer at $r = 0$ and with radius

$$R_{PH}(t) = a(t) \int_0^t \frac{dt'}{a(t')}. \tag{3.63}$$

The particle horizon contains every particle signal that has reached the observer between the time of the Big Bang $t = 0$ and the time t.[6]

For the fastest particles which travel radially to an observer at light speed, it is $ds^2 = 0$ and $d\Omega_{(2)}^2 = 0$. We write the line element using hyperspherical coordinates as in Eq. (3.10). Then, along ingoing radial null geodesics, it is $d\chi = -dt/a$ and the infinitesimal proper radius is $a(t)d\chi$. Integrating along an ingoing radial null geodesic between the emission of the signal at χ_e at time t_e and its detection at $\chi = 0$ at time t, one obtains

$$\int_{\chi_e}^0 d\chi = -\int_{t_e}^t \frac{dt'}{a(t')} \tag{3.64}$$

and the use of

$$\chi_e = \int_0^{\chi_e} d\chi = -\int_{\chi_e}^0 d\chi$$

leads to

$$\chi_e = \int_{t_e}^t \frac{dt'}{a(t')}. \tag{3.65}$$

---

[6]More realistically, photons propagate freely in the universe only after the time of the last scattering or recombination, before which the Compton scattering due to free electrons in the cosmic plasma makes it opaque. Therefore, cosmologists introduce the *optical horizon* with radius $a(t) \int_{t_{\text{recombination}}}^t \frac{dt'}{a(t')}$ [66]. However, the optical horizon is irrelevant for our purposes and will not be used here.

To obtain a physical (proper) radius R one multiplies by the scale factor obtaining[7]

$$R_e = a(t) \int_{t_e}^{t} \frac{dt'}{a(t')}, \tag{3.66}$$

Now take the limit $t_e \to 0^+$:

- If the integral $\int_0^t \frac{dt'}{a(t')}$ diverges, it is possible for the observer at $r = 0$ to receive all the light signals emitted at sufficiently early times from any point in the universe. The maximal volume that can be causally connected to the observer at time t is infinite.

- If the integral $\int_0^t \frac{dt'}{a(t')}$ is finite, the observer at $r = 0$ receives, at time t, only the light signals started within the sphere $r \leq \int_0^t \frac{dt'}{a(t')}$, which is called the (comoving) radius of the particle horizon.

The physical (proper) radius of the particle horizon is therefore given by Eq. (3.63). *At a given time t* the particle horizon is the boundary between the worldlines that can be seen by the observer and those ("beyond the horizon") which cannot be seen. This boundary hides events which cannot be known by that observer *at time t*. The particle horizon evolves with time (see Fig. 3.1).

The particle horizon is the horizon commonly studied in inflationary cosmology. The horizon problem of standard Big Bang cosmology consists of the fact that cosmic microwave photons coming from two opposite directions in the sky have the same temperature $T \simeq 2.73\,\mathrm{K}$ (apart from small fluctuations $\delta T/T \sim 5 \cdot 10^{-5}$) even though they come from regions which have never been in causal contact and cannot have thermalized. In the context of standard Big Bang theory this property can only be explained by invoking miracolously fine-tuned initial conditions. Inflation in the early universe provides a solution of the horizon problem by postulating that a small region of the universe has inflated at speeds much larger than those possible in the standard Big Bang model and that the cosmic microwave background photons that we see today were in causal contact in this small region before inflation occurred [42, 54, 57, 66].

Note that the particle and event horizons *depend on the observer*. Contrary to the case of the event horizon associated with a Schwarzschild black hole (often called an *absolute* horizon), different comoving observers in FLRW space will see event horizons located at different places (e.g., [26]). Another difference with respect to a black hole horizon is that the observer is located *inside* the cosmological particle horizon and signals sent *from the outside* cannot reach him or her. In this sense, the

---

[7]The notation for the proper radius $R \equiv a(t)r = a(t)f(\chi)$ is consistent with our previous use of this symbol to denote an areal radius because $a(t)r$ is in fact an areal radius, as is obvious from the inspection of the FLRW line element (3.10). If $k \neq 0$, the proper radius $a(t)\chi$ and the areal radius $a(t)f(\chi)$ do not coincide.

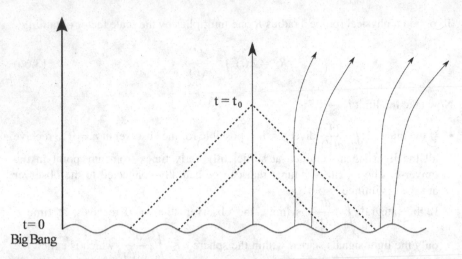

Fig. 3.1 The particle horizon of a comoving FLRW observer at time t. As the comoving observer moves along his or her (*vertical*) worldline, the particle horizon becomes larger and larger and this observer is able to see more and more signals coming from further and further away. Other observers (represented by the *curved* wordlines) will have different particle horizons

cosmological horizon of an expanding FLRW space resembles more a white hole than a black hole event horizon.

The cosmological particle horizon is a null surface.

Proof. This statement should be obvious from the fact that the event horizon is a causal boundary and is generated by the null geodesics which barely fail to reach the observer, but let us check it explicitly anyway. Using hyperspherical coordinates (t, χ), the equation of the particle horizon is

$$\mathscr{F}(t, \chi) \equiv \chi - \int_0^t \frac{dt'}{a(t')} = 0. \tag{3.67}$$

The normal to this surface has components

$$N_\mu = \nabla_\mu \mathscr{F} \,|_{\text{PH}} = \delta_{\mu 1} - \frac{\delta_{\mu 0}}{a} \tag{3.68}$$

and the norm squared of this normal is

$$N^a N_a = g^{ab} N_a N_b = g^{\mu\nu} \left(-\frac{\delta_{\mu 0}}{a} + \delta_{\mu 1} \right) \left(-\frac{\delta_{\nu 0}}{a} + \delta_{\nu 1} \right)$$

$$= \frac{g^{00}}{a^2} + g^{11} = -\frac{1}{a^2} + \frac{1}{a^2} = 0 \,,$$

hence the particle horizon is a null surface.   □

The cosmological particle horizon evolves according to the equation [24, 35]

$$\dot{R}_{PH} = H R_{PH} + 1 \,. \tag{3.69}$$

Proof. Equation (3.63) gives

$$\dot{R}_{PH} = \dot{a} \int_0^t \frac{dt'}{a(t')} + \frac{a}{a} = \frac{\dot{a}}{a} a(t) \int_0^t \frac{dt'}{a(t')} + 1 = H R_{PH} + 1 \,.$$

□

In an expanding universe with a particle horizon it is $\dot{R}_{PH} > 0$, which means that more and more signals emitted between the Big Bang and the time t reach the observer as the time t progresses. If $R_{PH}(t)$ does not diverge as $t \to t_{max}$ (where t_{max} is possibly $+\infty$), then there will always be a region unaccessible to the comoving observers.

The acceleration of the particle horizon is

$$\ddot{R}_{PH} = \left(\dot{H} + H^2 \right) R_{PH} + H = \frac{\ddot{a}}{a} R_{PH} + H \doteq -\frac{4\pi}{3} \left(\rho + 3P \right) R_{PH} + H \,, \tag{3.70}$$

as follows from Eqs. (3.69) and (3.6).

3.8.2 Event Horizon

Consider now all the events which can be seen by a comoving observer at $r = 0$ between time t and future infinity $t = +\infty$ (in a closed universe which recollapses, or in a Big Rip universe which ends at a finite time, substitute $+\infty$ with the time t_{max} corresponding to the maximal expansion or the Big Rip, respectively). The comoving radius of the region which can be seen by this observer is

$$\chi_{EH} = \int_t^{+\infty} \frac{dt'}{a(t')} \,; \tag{3.71}$$

if this integral diverges as the upper limit of integration goes to infinity or to t_{max}, it is said that there is no event horizon in this FLRW space and events arbitrarily far away can eventually be seen by the observer by waiting a sufficiently long time. If the integral converges, there is an *event horizon*: events beyond r_{EH} will *never* be known to the observer [75]. The physical (proper) radius of the event horizon is

$$R_{EH}(t) = a(t) \int_t^{+\infty} \frac{dt'}{a(t')} \,. \tag{3.72}$$

In short, the event horizon can be said to be the "complement" of the particle horizon [66]; it is the (proper) distance to the most distant event that the comoving observer will ever see. Clearly, in order to define the event horizon one must know the entire future history of the universe from time t to infinity and the event horizon is defined *globally*, not locally.

The cosmological event horizon is a null surface.

Proof. Again, the statement follows from the fact that the event horizon is a causal boundary. To check explicitly, the equation of the particle horizon in hyperspherical coordinates (t, χ) is

$$\mathscr{F}(t, \chi) \equiv \chi - \int_t^{t_{\max}} \frac{dt'}{a(t')} = 0. \tag{3.73}$$

The normal to this surface has components

$$N_\mu = \nabla_\mu \mathscr{F} \mid_{\text{EH}} = \delta_{\mu 1} - \frac{\delta_{\mu 0}}{a} \tag{3.74}$$

and the norm squared of this normal is

$$N^a N_a = g^{ab} N_a N_b = g^{\mu\nu} \left(\delta_{\mu 1} - \frac{\delta_{\mu 0}}{a} \right) \left(\delta_{\nu 1} - \frac{\delta_{\nu 0}}{a} \right)$$

$$= g^{11} + \frac{g^{00}}{a^2} = \frac{1}{a^2} - \frac{1}{a^2} = 0. \tag{3.75}$$

□

The cosmological event horizon evolves according to the equation [1, 24, 35, 64]

$$\dot{R}_{\text{EH}} = H R_{\text{EH}} - 1. \tag{3.76}$$

Proof. Equation (3.72) gives

$$\dot{R}_{\text{EH}} = \dot{a} \int_t^{t_{\max}} \frac{dt'}{a(t')} + a(t) \frac{d}{dt} \left(-\int_{t_{\max}}^t \frac{dt'}{a(t')} \right) = Ha(t) \int_t^{t_{\max}} \frac{dt'}{a(t')} - \frac{a}{a}$$

$$= H R_{\text{EH}} - 1.$$

□

The acceleration of the event horizon is also straightforward to derive,

$$\ddot{R}_{\text{EH}} = \left(\dot{H} + H^2 \right) R_{\text{EH}} - H. \tag{3.77}$$

The event horizon does not exist in every FLRW space.

To wit, consider a spatially flat ($k = 0$) FLRW universe sourced by a perfect fluid with equation of state $P = w\rho$ and $w = $ constant > -1. If $w > -1/3$ (i.e., in General Relativity, for a decelerating universe[8]), there is no event horizon because

$$a(t) = a_0 \, t^{\frac{2}{3(w+1)}} \qquad (3.78)$$

and the event horizon radius would be

$$R_{EH} = a_0 \, t^{\frac{2}{3(w+1)}} \int_t^{+\infty} \frac{dt'}{a_0 \, (t')^{\frac{2}{3(w+1)}}} = t^{\frac{2}{3(w+1)}} \left[\frac{3(w+1)}{3w+1} \, t'^{\frac{3w+1}{3(w+1)}} \right]_t^{+\infty} .$$

If $w > -1/3$ then the exponent $\dfrac{3w+1}{3(w+1)}$ is positive and the integral diverges: there is no event horizon in this case. Indeed, the existence of cosmological event horizons seems to require the violation of the strong energy condition in at least some region of spacetime [10]. We can state that

in General Relativity with a perfect fluid the FLRW event horizon exists only for accelerated universes with $P < -\rho/3$.

Since conformal time η is defined by $dt = a d\eta$, it is

$$R_{EH}(\eta) = a(\eta) \, (\eta_{max} - \eta) .$$

For a spatially flat universe with perfect fluid, $P = w\rho$ and $w = $ constant $\neq -1$ in General Relativity, using Eq. (3.78) one obtains

$$\eta = \frac{3(w+1)}{(3w+1)a_0} \, t^{\frac{3w+1}{3(w+1)}} . \qquad (3.79)$$

If $-1 < w < -1/3$, then η is negative for $t > 0$ while, if $w < -1$ (phantom universe) or if $w > -1/3$ (decelerating universe), it is $\eta > 0$ for $t > 0$.

For $w = -1$ we have de Sitter space with scale factor $a(t) = a_0 \, e^{Ht}$ (a_0, H constants) and $dt \equiv a d\eta = a_0 e^{Ht} d\eta$, which yields

$$\eta = \int d\eta = \int dt \, \frac{e^{-Ht}}{a_0} = -\frac{1}{aH} = -\frac{e^{-Ht}}{a_0 H} < 0 .$$

Then, $t \to -\infty$ corresponds to $\eta \to -\infty$, $t \to +\infty$ to $\eta \to 0^-$, and we have $t = -H^{-1} \ln (a_0 H |\eta|)$.

---

[8]The Einstein-Friedmann equation (3.6) gives $\ddot{a} < 0$ in this case.

3.8.3   Hubble Horizon

For completeness, we mention that the literature refers often to a *Hubble horizon* with radius

$$R_{\mathrm{H}} \equiv \frac{1}{H}. \tag{3.80}$$

This is just language: this equation is not derived from any particular physical consideration, other than giving an order of magnitude of the radius of curvature of a FLRW space and being used as an estimate of the radius of the event horizon during slow-roll inflation, when the universe is close to a de Sitter space [54]. The Hubble horizon coincides with the apparent horizon for spatially flat universes and with the horizon of de Sitter space. However, the concept of Hubble horizon does not add to the discussion of the physics of the various types of FLRW horizons and it will not be used in the following.

3.8.4   Apparent Horizon

According to the definition (2.57)–(2.58), the apparent horizon of FLRW space is located by setting to zero the expansion θ_n given by Eq. (3.53) while θ_l (given by Eq. (3.52)) is positive. Alternatively, note that the FLRW metric is obviously spherically symmetric and one can apply the prescription (2.104) to locate the apparent horizon deriving from the Misner-Sharp-Hernandez mass, as we have already done. Using the areal radius R, the FLRW line element reads

$$ds^2 = -dt^2 + \frac{a^2(t)}{1 - kr^2} \, dr^2 + R^2 d\Omega_{(2)}^2 \tag{3.81}$$

where $r(t, R) = R/a(t)$, which is of the form (2.91) with

$$h_{ab} = \mathrm{diag}\left(-1, \frac{a^2(t)}{1 - kr^2}, 0, 0\right) \tag{3.82}$$

in comoving coordinates. The apparent horizon is located by the equation $\nabla^c R \nabla_c R = 0$, or

$$-R_t^2 + \left(\frac{1 - kr^2}{a^2}\right) R_r^2 = 0. \tag{3.83}$$

Since $R_t = \dot{a}(t)r$ and $R_r = a(t)$, we obtain $\dot{a}^2 r^2 = 1 - kr^2$ or $H^2 R^2 = 1 - kR^2/a^2$, and the *FLRW apparent horizon* of a comoving observer is a sphere of proper radius

$$R_{\mathrm{AH}}(t) = \frac{1}{\sqrt{H^2 + k/a^2}} \tag{3.84}$$

centered on this observer. Looking at Eqs. (3.52) and (3.53) it is clear that when $R > R_{AH}$ it is $\theta_l > 0$ and $\theta_n > 0$, while the region $0 \le R < R_{AH}$ has $\theta_l > 0$ and $\theta_n < 0$ (radial null rays coming from the region outside the horizon will not cross it and reach the observer). The cosmological apparent horizon depends on the observer (here chosen at $r = 0$), much like horizons in flat space. It acts as a sphere surrounding the observer and hiding information.

For a spatially flat universe, the radius of the apparent horizon R_{AH} coincides with the Hubble radius H^{-1} while, for a positively curved ($k > 0$) universe, R_{AH} is smaller than the Hubble radius, and it is larger for an open ($k < 0$) universe. Note that, in General Relativity, the Hamiltonian constraint (3.5) guarantees that the argument of the square root in Eq. (3.84) is positive for positive densities ρ. The apparent horizon always exists (except for the trivial case of Minkowski spacetime without gravity) while, as seen above, the event horizon does not exist in all FLRW spaces.

The apparent horizon, in general, is not a null surface, contrary to the event and particle horizons.

To prove this statement, note that the equation of the apparent horizon in comoving coordinates is

$$\mathcal{F}(t,r) = a(t)r - \frac{1}{\sqrt{H^2 + k/a^2}} = 0 . \tag{3.85}$$

The normal to this surface has components

$$
\begin{aligned}
N_\mu = \nabla_\mu \mathcal{F} \, |_{AH} &= \left[\dot{a}r\,\delta_{\mu 0} + a\,\delta_{\mu 1} + \frac{H\dot{H}\,\delta_{\mu 0} - \frac{k\dot{a}}{a^3}\,\delta_{\mu 0}}{(H^2 + k/a^2)^{3/2}} \right]_{AH} \\
&= \left\{ \left[\dot{a}r + \frac{H\left(\dot{H} - k/a^2\right)}{(H^2 + k/a^2)^{3/2}} \right]\delta_{\mu 0} + a\,\delta_{\mu 1} \right\}_{AH} \\
&= \left[HR_{AH} + \frac{H\left(\dot{H} - k/a^2\right)}{(H^2 + k/a^2)^{3/2}} \right]\delta_{\mu 0} + a\,\delta_{\mu 1} \\
&= HR_{AH} \left[1 + \left(\dot{H} - \frac{k}{a^2}\right)R_{AH}^2 \right]\delta_{\mu 0} + a\,\delta_{\mu 1} \\
&= HR_{AH}^3 \frac{\ddot{a}}{a}\,\delta_{\mu 0} + a\,\delta_{\mu 1} .
\end{aligned}
\tag{3.86}
$$

In General Relativity with a perfect fluid with equation of state $P = w\rho$, Eqs. (3.5), (3.6), and (3.84) yield

$$N_\mu \doteq -\frac{(3w+1)}{2}\, HR_{AH}\delta_{\mu 0} + a\delta_{\mu 1} . \tag{3.87}$$

The norm squared of the normal is

$$N^a N_a = g^{ab} N_a N_b$$

$$= g^{\mu\nu} \left[\left(HR_{AH} + \frac{H\left(\dot{H} - k/a^2\right)}{(H^2 + k/a^2)^{3/2}} \right) \delta_{\mu 0} + a\, \delta_{\mu 1} \right] \cdot$$

$$\cdot \left[\left(HR_{AH} + \frac{H\left(\dot{H} - k/a^2\right)}{(H^2 + k/a^2)^{3/2}} \right) \delta_{\nu 0} + a\, \delta_{\nu 1} \right]$$

$$= g^{\mu\nu} \left\{ \left[HR_{AH} + \frac{H\left(\dot{H} - k/a^2\right)}{(H^2 + k/a^2)^{3/2}} \right]^2 \delta_{\mu 0}\delta_{\nu 0} \right.$$

$$\left. + \cdots \delta_{\mu 1}\delta_{\nu 0} + \cdots \delta_{\mu 0}\delta_{\nu 1} + a^2\, \delta_{\mu 1}\delta_{\nu 1} \right\}$$

$$= g^{00} H^2 \left[R_{AH} + \frac{\left(\dot{H} - k/a^2\right)}{(H^2 + k/a^2)^{3/2}} \right]^2 + a^2 g^{11}$$

$$= 1 - kr_{AH}^2 - H^2 \left[R_{AH} + \frac{\left(\dot{H} - k/a^2\right)}{(H^2 + k/a^2)^{3/2}} \right]^2$$

$$= 1 - kr_{AH}^2 - H^2 \left[\frac{1}{\sqrt{H^2 + k/a^2}} + \frac{\left(\dot{H} - k/a^2\right)}{(H^2 + k/a^2)^{3/2}} \right]^2$$

$$= 1 - kr_{AH}^2 - \frac{H^2 \left(\dot{H} + H^2\right)^2}{(H^2 + k/a^2)^3} = 1 - kr_{AH}^2 - \frac{H^2 \left(\ddot{a}/a\right)^2}{H^6 \left(1 + \frac{k}{a^2 H^2}\right)^3}$$

$$= H^2 R_{AH}^2 \left[1 - \left(\frac{\ddot{a}}{a}\right)^2 R_{AH}^4 \right]$$

$$\doteq \frac{3 H^2 R_{AH}^2}{4\rho^2} \left(\rho + P\right)\left(\rho - 3P\right)$$

$$= H^2 R_{AH}^2 \left(1 - q^2 H^4 R_{AH}^4\right) , \qquad (3.88)$$

where $q \equiv -\ddot{a}a/\dot{a}^2$ is the deceleration parameter. The horizon is null if and only if $P = -\rho$ or if $P = \rho/3$.

In General Relativity with a perfect fluid the Hamiltonian constraint (3.5) yields

$$H^2 R_{AH}^2 = \left(1 + \frac{k}{a^2 H^2}\right)^{-1} = \left(\frac{8\pi}{3H^2}\rho\right)^{-1} \equiv \frac{\rho_c}{\rho} \equiv \Omega^{-1}, \qquad (3.89)$$

where $\rho_c \equiv \dfrac{3H^2}{8\pi}$ is the critical density and $\Omega \equiv \rho/\rho_c$ is the density parameter, i.e., the energy density measured in units of the critical density, and one obtains

$$N^a N_a \doteq -\frac{3}{4}(w+1)(3w-1)H^2 R_{\mathrm{AH}}^2 \doteq \frac{\Omega^2 - q^2}{\Omega^3} . \tag{3.90}$$

We can use Eq. (3.90) to establish that [9, 12, 35, 77]:

- If $-1 < w < 1/3$ then $N^c N_c > 0$ and the apparent horizon is *timelike*. For a $k = 0$ universe in Einstein theory this condition corresponds to $\dot{H} < 0$.
- If $w = -1$ or $w = 1/3$ then $N^c N_c = 0$ and the apparent horizon is *null* (de Sitter space, which has $\dot{H} = 0$ and $q = -1$, falls into this category but it is not the only space with this horizon property).
- If $w < -1$ or $w > 1/3$ then $N^c N_c < 0$, the normal is timelike, and the apparent horizon is *spacelike*. In Einstein theory with $k = 0$ and a perfect fluid as the source, $w < -1$ corresponds to $\dot{H} > 0$ ("superacceleration"). This is the case of Big Rip universes and of a phantom fluid which violates the weak energy condition.

The black hole dynamical horizons discussed in the literature are usually required to be spacelike [4]. However, cosmological horizons in the presence of non-exotic matter are timelike.

In General Relativity, the radius of the apparent horizon can be written as

$$R_{\mathrm{AH}} = \frac{1}{\sqrt{\Omega}\,|H|} \tag{3.91}$$

in terms of the density parameter Ω by using Eq. (3.89).

The apparent horizon evolves according to the equation [2, 19, 21, 24, 35, 60]

$$\dot{R}_{\mathrm{AH}} = HR_{\mathrm{AH}}^3 \left(\frac{k}{a^2} - \dot{H} \right) \doteq 4\pi HR_{\mathrm{AH}}^3 (P + \rho) . \tag{3.92}$$

Proof. Differentiate Eq. (3.84) with respect to t, obtaining

$$\dot{R}_{\mathrm{AH}} = -\left[\frac{H\left(\dot{H} - k/a^2\right)}{\left(H^2 + k/a^2\right)^{3/2}} \right]_{\mathrm{AH}}$$

$$= H\left(-\dot{H} + \frac{k}{a^2}\right) R_{\mathrm{AH}}^3$$

$$= HR_{\mathrm{AH}}^3 \left(\frac{1}{R_{\mathrm{AH}}^2} - H^2 - \dot{H} \right)$$

$$= HR_{\mathrm{AH}} \left[1 - \left(\dot{H} + H^2\right) R_{\mathrm{AH}}^2 \right]$$

$$= HR_{\mathrm{AH}} \left(1 - \dot{H}R_{\mathrm{AH}}^2 - \frac{H^2}{H^2 + k/a^2} \right) = HR_{\mathrm{AH}}^3 \left(\frac{k}{a^2} - \dot{H} \right) .$$

$\square$

In General Relativity with a perfect fluid as a source, the only way to obtain a stationary apparent horizon is when $P = -\rho$. This equation of state yields de Sitter space, for which Eq. (3.92) reduces to $\dot{R}_{AH} = 0$, consistent with $R_H = H^{-1}$ and $H = $ constant. For non-spatially flat universes, the equation of state $P = -\rho$ produces other solutions.

Example 3.1. For $k = -1$ and a cosmological constant $\Lambda > 0$ as the only source of gravity, the scale factor

$$a(t) = \sqrt{\frac{3}{\Lambda}} \sinh\left(\sqrt{\frac{\Lambda}{3}} t\right) \qquad (3.93)$$

is a solution of the Einstein-Friedmann equations, as can be checked easily. The radius of the event horizon has the time dependence

$$R_{EH}(t) = \sqrt{\frac{3}{\Lambda}} \sinh\left(\sqrt{\frac{\Lambda}{3}} t\right) \left| \ln\left[\tanh\left(\sqrt{\frac{\Lambda}{3}} \frac{t}{2}\right)\right]\right| . \qquad (3.94)$$

The apparent horizon, instead, has constant radius $R_{AH} = \sqrt{3/\Lambda}$, according to the fact that $\rho_\Lambda + P_\Lambda = 0$ in Eq. (3.92) [35].

Example 3.2. Consider, for $k = +1$ and positive cosmological constant Λ, the scale factor

$$a(t) = \sqrt{\frac{3}{\Lambda}} \cosh\left(\sqrt{\frac{\Lambda}{3}} t\right) ; \qquad (3.95)$$

the event horizon has radius

$$R_{EH}(t) = \sqrt{\frac{3}{\Lambda}} \cosh\left(\sqrt{\frac{\Lambda}{3}} t\right) \left[\frac{\pi}{2} + n\pi - \tan^{-1}\left(\sinh\left(\sqrt{\frac{\Lambda}{3}} t\right)\right)\right] , \qquad (3.96)$$

where $n = 0, \pm 1, \pm 2, \ldots$ The multiple possible values of n correspond to the infinite possible branches which one can consider when inverting the tangent function, and to the fact that in a closed universe light rays can travel multiple times around the universe. In this situation it is problematic to regard the event horizon as a true horizon [26]. The apparent horizon has constant radius $R_{AH} = \sqrt{3/\Lambda}$, according to the fact that $\rho_\Lambda + P_\Lambda = 0$ in Eq. (3.92).[9]

---

[9]These two examples, together with de Sitter space for $k = 0$, are presented in Ref. [26]. However, contrary to what is stated in this reference, in both cases the event horizon is not constant: it is the apparent horizon which is constant.

3.8.5 Trapping Horizon

When is the FLRW apparent horizon also a trapping horizon? When $\mathcal{L}_l \theta_n > 0$, which gives

$$\mathcal{L}_l \theta_n = \frac{\mathcal{R}}{3} > 0, \tag{3.97}$$

where $\mathcal{R}$ is the Ricci scalar of FLRW space [8, 9, 12, 35, 77].

Proof. Using Eqs. (3.48) and (3.53), we have

$$\mathcal{L}_l \theta_n = l^a \nabla_a \theta_n = l^a \partial_a \theta_n$$

$$= 2 \left(\partial_t + \frac{\sqrt{1 - kr^2}}{a} \partial_r \right) \left(H - \frac{\sqrt{1 - kr^2}}{ar} \right)$$

$$= \frac{2}{R^2} \left(\dot{H} R^2 + H R \sqrt{1 - \frac{kR^2}{a^2}} + 1 \right).$$

At the apparent horizon $R = R_{\mathrm{AH}}$ it is

$$\mathcal{L}_l \theta_n \big|_{\mathrm{AH}} = 2 \left(H^2 + \frac{k}{a^2} \right) \left[\frac{\dot{H}}{H^2 + \frac{k}{a^2}} + \frac{H \sqrt{1 - \frac{k}{a^2(H^2 + k/a^2)}}}{\sqrt{H^2 + k/a^2}} + 1 \right]$$

$$= 2 \left(\dot{H} + 2H^2 + \frac{k}{a^2} \right) = \frac{\mathcal{R}}{3}.$$

$\square$

In General Relativity with a perfect fluid it is

$$\mathcal{L}_l \theta_n \big|_{\mathrm{AH}} \doteq \frac{8\pi}{3} (\rho - 3P) \tag{3.98}$$

and, therefore [8, 9, 12, 35, 77]

the apparent horizon is also a trapping horizon iff $\mathcal{R} > 0$ (equivalent to $P < \rho/3$ in General Relativity with a perfect fluid).

When trapping, the apparent horizon is a past inner trapping horizon according to Hayward's definition [45]: it is an *inner* horizon because the region $0 < R < R_{\mathrm{AH}}(t)$ is not trapped (i.e., $\theta_l \theta_n < 0$), and a *past* horizon because geodesics which have exited the hypersurface $R = R_{\mathrm{AH}}$ cannot come back to it.

3.8.6 Examples

Let us consider a couple of examples.

Example 3.3. As a first example consider Minkowski space with line element

$$ds^2 = -dt^2 + dR^2 + R^2 d\Omega^2_{(2)} : \tag{3.99}$$

this is a trivial $k = 0$ FLRW universe with constant scale factor $a \equiv 1$. The expression (3.84) of the radius of the apparent horizon gives $R_{\rm AH} = \infty$, i.e., the apparent horizon is located at spatial infinity.

Example 3.4. As a second example consider the *Milne universe* with line element

$$ds^2 = -d\tau^2 + \tau^2 \left(d\chi^2 + \sinh^2 \chi \, d\Omega^2_{(2)} \right) , \tag{3.100}$$

which describes a $k = -1$ FLRW universe in hyperspherical coordinates with linear scale factor $a(\tau) = \tau$. The areal radius is $R = \tau \sinh \chi$ and its gradient has components

$$\nabla_\mu R = \sinh \chi \, \delta_{\mu 0} + \tau \cosh \chi \, \delta_{\mu 1} ,$$

which gives

$$\nabla^c R \nabla_c R = 1 ; \tag{3.101}$$

since $\nabla_c R$ cannot be null there is no apparent horizon. This fact is consistent with Eq. (3.84) which yields $R_{\rm AH} = +\infty$. This fact is not surprising because the Milne universe is nothing but a portion of empty Minkowski space. In fact, for $a = \tau$ and $k = -1$, the Einstein-Friedmann equations with a perfect fluid (3.5) and (3.6) give $\rho = P = 0$: this FLRW universe is empty and must therefore have zero spacetime curvature and be Minkowski space, which is confirmed by a direct calculation of the Riemann tensor. The Milne line element can be obtained from the Minkowski line element (3.99) with the coordinate transformation $(t, R, \theta, \varphi) \rightarrow (\tau, \chi, \theta, \varphi)$ given by

$$t = \tau \cosh \chi , \tag{3.102}$$

$$R = \tau \sinh \chi , \tag{3.103}$$

as is straightforward to verify. The Milne universe is a foliation of flat Minkowski spacetime with curved three-dimensional hypersurfaces. Although, in general, the existence of apparent horizons is foliation-dependent, in this case changing the foliation does not introduce an apparent horizon in Minkowski spacetime. The dependence on the spacetime slicing does not always lead to trouble.

3.9 Dynamics of Cosmological Horizons

Let us now focus on the dynamical evolution of the horizons that we have defined and compare their evolutionary laws, which we summarize here. The first question to ask is whether these horizons are comoving: they almost never are. The difference between the expansion rate $\dot{R}/R$ of a horizon of radius R and the expansion rate H of the cosmic matter is, for the particle, event, and apparent horizons

$$\frac{\dot{R}_{PH}}{R_{PH}} - H = \frac{1}{R_{PH}}, \tag{3.104}$$

$$\frac{\dot{R}_{EH}}{R_{EH}} - H = -\frac{1}{R_{EH}}, \tag{3.105}$$

$$\frac{\dot{R}_{AH}}{R_{AH}} - H = H\left[\frac{\left(\frac{k}{a^2} - \dot{H}\right)}{H^2 + \frac{k}{a^2}} - 1\right] = -\left(\frac{\ddot{a}}{a}H\right)R_{AH}^2 \doteq \frac{(3w+1)H}{2}, \tag{3.106}$$

respectively. Taking into consideration only expanding FLRW universes ($H > 0$), when it exists the particle horizon always expands faster than comoving. The event horizon (which only exists for accelerated universes) always expands slower than comoving. The apparent horizon expands faster than comoving for decelerated universes ($\ddot{a} < 0$); slower than comoving for accelerated universes ($\ddot{a} > 0$); and comoving for coasting universes ($a(t) \propto t$).

An even simpler way of looking at the evolution is by using the comoving radius of the horizon: if this radius is constant, then the horizon is comoving. We have,

$$\dot{r}_{PH} = \frac{1}{a} > 0, \tag{3.107}$$

$$\dot{r}_{EH} = -\frac{1}{a} < 0, \tag{3.108}$$

$$\dot{r}_{AH} = -\frac{\dot{a}\ddot{a}}{(\dot{a}^2 + k)^{3/2}}, \tag{3.109}$$

respectively. All these horizons are practically never comoving.

In a universe which changes its expansion from accelerated to decelerated, the event horizon suddenly disappears when $\ddot{a}(t)$ changes sign. Vice-versa, an event horizon appears when the universe goes from decelerated to accelerated. If an event horizon was to be assigned an entropy equal to one quarter of its area (which is not an established fact), then this behaviour causes discontinuous jumps in the horizon entropy. In a de Sitter universe the horizon is, instead, static and so is the entropy.

3.10   Another Notation

The notation and terminology may become confusing when switching from black
hole to cosmological horizons and from observers outside an horizon to observers
surrounded by a horizon. It is perhaps more convenient to adopt another common
notation: instead of using l^a and n^a for the tangents to the outgoing and ingoing null
geodesic congruences in a black hole spacetime and having to invert these when
discussing white hole or expanding FLRW spacetimes, one can simply use $l^a_{(\pm)}$
where "+" and "−" denote outgoing or ingoing congruences as appropriate. The
terminology is as follows.[10]

- For a *black hole* + denotes the outgoing null congruence (with tangent l^a in our
  previous notation) and − denotes the ingoing null congruence (with tangent n^a in
  the other notation). The observer is outside the apparent horizon.

  - A surface is *normal* (untrapped) if $\theta_+ > 0$ and $\theta_- < 0$.
  - A surface is *trapped* if $\theta_+ < 0$ and $\theta_- < 0$.
  - A surface is *(future) marginally trapped* if $\theta_+ = 0$ and $\theta_- < 0$ and is *outer* if
    $\partial_- \theta_+ < 0$.

- For a *white hole* + denotes the outgoing and − the ingoing null congruences.
  The observer is outside the apparent horizon.

  - A surface is *normal* (untrapped) if $\theta_+ > 0$ and $\theta_- < 0$.
  - A surface is *trapped* if $\theta_+ > 0$ and $\theta_- > 0$.
  - A surface is *(past) marginally trapped* if $\theta_+ > 0$ and $\theta_- = 0$ and is *outer* if
    $\partial_+ \theta_- < 0$.

- For an *expanding FLRW space* + denotes the outgoing and − the ingoing null
  congruences. The observer is inside the apparent horizon.

  - A surface is *normal* (untrapped) if $\theta_+ > 0$ and $\theta_- < 0$.
  - A surface is *trapped* if $\theta_+ > 0$ and $\theta_- > 0$.
  - A surface is *(past) marginally trapped* if $\theta_+ > 0$ and $\theta_- = 0$ and is *inner* if
    $\partial_+ \theta_- > 0$.

- For a *contracting FLRW space* + denotes the outgoing and − the ingoing null
  congruences. The observer is inside the apparent horizon.

  - A surface is *normal* (untrapped) if $\theta_+ > 0$ and $\theta_- < 0$.
  - A surface is *trapped* if $\theta_+ < 0$ and $\theta_- < 0$.
  - A surface is *(future) marginally trapped* if $\theta_+ = 0$ and $\theta_- < 0$ and is *inner* if
    $\partial_- \theta_+ > 0$.

---

[10]Cf. Table 1 of Ref. [39].

3.11 de Sitter Space

de Sitter space (see Ref. [44] for an introduction and Refs. [6, 16, 78, 79] for reviews)
is a maximally symmetric constant curvature space and is a spatially flat FLRW
space with line element

$$ds^2 = -dt^2 + a_0^2 \, e^{2Ht} \left(dr^2 + r^2 d\Omega_{(2)}^2 \right) \qquad (3.110)$$

in spherical comoving coordinates. The scale factor is $a(t) = a_0 \, e^{Ht}$, with a_0 and
H constants. The de Sitter solution is obtained from the Einstein-Friedmann equations (3.5) and (3.6) with a positive cosmological constant Λ (which can formally
be treated as a perfect fluid with $\rho = -P = \dfrac{\Lambda}{8\pi}$) and no matter, and $H = \pm\sqrt{\Lambda/3}$.
Sometimes the literature refers to *expanding de Sitter spaces* with $a(t) = a_0 e^{\sqrt{\Lambda/3}\,t}$
and to *contracting de Sitter spaces* with $a(t) = a_0 e^{-\sqrt{\Lambda/3}\,t}$. Other times the scale
factor $a(t) = a_0 \cosh\left(\sqrt{\Lambda/3}\,t \right)$ is used which is a combination of the previous two.
We will call "de Sitter space" only the one with line element (3.110) and scale
factor $a(t) = a_0 e^{Ht}$, with H a *positive* constant. It has curvature index $k = 0$ and
Ricci scalar $\mathscr{R} = 12H^2$.

The de Sitter metric can be recast in static Schwarzschild-like coordinates as
follows. First, introduce the areal radius $R(t, r) \equiv a(t)r$, in terms of which

$$dr = \frac{dR}{a} - \frac{R\dot{a}}{a^2}\, dt \, ; \qquad (3.111)$$

this change of radius reduces the line element to the *Painlevé-Gullstrand form*[11]

$$ds^2 = -\left(1 - H^2 R^2\right) dt^2 - 2HR\, dtdR + dR^2 + R^2 d\Omega_{(2)}^2 \, . \qquad (3.112)$$

The Painlevé-Gullstrand coordinates penetrate the horizon $R = 1/H$ and constant
time slices are Euclidean (by setting $dt = 0$ one obtains the 3-dimensional Euclidean
line element $ds^2|_{(3)} = dR^2 + R^2 d\Omega_{(2)}^2$), properties which we have already seen to
hold for the Painlevé-Gullstrand coordinates of the Schwarzschild metric.

The cross-term proportional to $dtdR$ in the line element (3.112) is eliminated by
the use of the new time coordinate T defined by

$$dT = dt + \frac{HR}{1 - H^2 R^2}\, dR \, . \qquad (3.113)$$

dT is a locally exact differential: in fact, thinking of (t, R) as independent coordinates, it is $dT = T_1 dt + T_2 dR$ with $T_1 = 1$, $T_2 = HR/(1 - H^2 R^2)$, and

---

[11]The coordinates (t, R, θ, φ) are called "Painlevé-de Sitter coordinates" [72].

$$\frac{\partial T_1}{\partial R} = \frac{\partial T_2}{\partial t} = 0 .$$

By substituting $dt = dT - \dfrac{HR}{1 - H^2 R^2}\, dR$ into the line element (3.112), one obtains

$$ds^2 = - \left(1 - H^2 R^2\right) \left(dT^2 - \frac{2HR}{1 - H^2 R^2}\, dTdR + \frac{H^2 R^2}{\left(1 - H^2 R^2\right)^2}\, dR^2 \right)$$

$$- 2HR \left(dT - \frac{HR}{1 - H^2 R^2}\, dR \right) dR + dR^2 + R^2 d\Omega^2_{(2)}$$

$$= - \left(1 - H^2 R^2\right) dT^2$$

$$+ \left[1 - \left(1 - H^2 R^2\right) \frac{H^2 R^2}{\left(1 - H^2 R^2\right)^2} + \frac{2H^2 R^2}{\left(1 - H^2 R^2\right)} \right] dR^2$$

$$+ R^2 d\Omega^2_{(2)} \tag{3.114}$$

and finally the *de Sitter metric in static Schwarzschild-like coordinates* is obtained (e.g., [41, 65]),

$$ds^2 = - \left(1 - H^2 R^2\right) dT^2 + \frac{dR^2}{1 - H^2 R^2} + R^2 d\Omega^2_{(2)} . \tag{3.115}$$

This line element bears some resemblance with the Schwarzschild metric in Schwarzschild coordinates. This static chart covers only the region $0 < R < R_{\text{EH}}$ of de Sitter space interior to the horizon and, therefore, de Sitter space is only locally static.

For $H = 0$ de Sitter space degenerates into Minkowski space, the static coordinate system becomes global, and there are no apparent and trapping horizons, $R_{\text{AH}} \to \infty$.

There is no particle horizon in de Sitter space (defined as the $k = 0$ FLRW space with scale factor $a(t) = a_0 e^{Ht}$ and $H > 0$) because there is no Big Bang and the integral

$$\int_{-\infty}^{t} \frac{dt'}{a(t')} = \left[\frac{-e^{-Ht'}}{a_0 H} \right]_{-\infty}^{t} \tag{3.116}$$

diverges.

There is an *event horizon in de Sitter space*, with radius

$$R_{\text{EH}} = \frac{1}{H} = \sqrt{\frac{3}{\Lambda}} , \tag{3.117}$$

which is time-independent.

Proof.

$$R_{\text{EH}} \equiv a(t) \int_t^{+\infty} \frac{dt'}{a(t')} = e^{Ht} \left[\frac{-e^{-Ht'}}{H} \right]_t^{+\infty} = \frac{1}{H} .$$

□

The area of the event horizon is $\mathscr{A}_{\text{EH}} = 4\pi R_{\text{EH}}^2 = 12\pi/\Lambda$.

Apparent and event horizons coincide in de Sitter space and this horizon is null.

This surface is also a Killing horizon of the Killing vector field

$$k^a \equiv \left(\frac{\partial}{\partial T} \right)^a . \tag{3.118}$$

To check that this is a Killing field, consider its components in static Schwarzschild-like coordinates

$$k^\mu = \left(1, 0, 0, 0 \right), \qquad k_\mu = \left(-(1 - H^2 R^2), 0, 0, 0 \right) \tag{3.119}$$

and the covariant derivatives

$$\nabla_\mu k_\nu = \partial_\mu k_\nu - \Gamma_{\mu\nu}^\alpha k_\alpha = \partial_\mu k_\nu + \Gamma_{\mu\nu}^0 \left(1 - H^2 R^2 \right) . \tag{3.120}$$

The only non-vanishing Christoffel symbols $\Gamma_{\mu\nu}^0$ in these coordinates are

$$\Gamma_{01}^0 = \Gamma_{10}^0 = -\frac{H^2 R}{1 - H^2 R^2} \tag{3.121}$$

or

$$\Gamma_{\mu\nu}^0 = -\frac{H^2 R}{1 - H^2 R^2} \left(\delta_{\mu 0} \delta_{\nu 1} + \delta_{\mu 1} \delta_{\nu 0} \right) ; \tag{3.122}$$

this equation in turn gives

$$\nabla_\mu k_\nu = H^2 R \left(\delta_{\mu 1} \delta_{\nu 0} - \delta_{\mu 0} \delta_{\nu 1} \right) \tag{3.123}$$

and the only component of the left hand side of the Killing equation $\nabla_a k_b + \nabla_b k_a = 0$ which is not trivially zero is

$$\nabla_0 k_1 + \nabla_1 k_0 = 2 \left(-H^2 R + H^2 R \right) , \tag{3.124}$$

which also vanishes. The Killing equation is satisfied and k^a is a Killing vector field with norm squared

$$k^a k_a = -\left(1 - H^2 R^2\right) . \tag{3.125}$$

Therefore, k^a is *timelike* for $R < H^{-1}$, *null* for $R = H^{-1}$, and *spacelike* for $R > H^{-1}$ and the hypersurface $R = H^{-1}$ is a Killing horizon. This situation is analogous to that of the Killing vector field $(\partial/\partial t)^a$ in the Schwarzschild spacetime (1.3), which changes its causal character on the Schwarzschild event horizon $R = 2M$.

The components of the de Sitter space Killing vector k^a in comoving coordinates are

$$k^\mu = (1, 0, 0, 0) , \qquad k_\mu = (-1, 0, 0, 0) . \tag{3.126}$$

In Painlevé-Gullstrand coordinates the components of the Killing field are instead

$$k^\mu = (1, \dot{a}r, 0, 0) = (1, HR, 0, 0) . \tag{3.127}$$

The surface gravity of the Killing horizon is

$$\kappa_{dS} = \frac{1}{R_{EH}} = H = \sqrt{\frac{\Lambda}{3}} . \tag{3.128}$$

Proof. The Killing surface gravity is given by Eq. (2.78) which yields, using static coordinates and Eq. (3.123),

$$\kappa_{Killing}^2 = -\frac{1}{2} \left(\nabla^a k^b\right) \left(\nabla_a k_b\right) = -\frac{1}{2} g^{\mu\alpha} g^{\nu\beta} \left(\nabla_\mu k_\nu\right) \left(\nabla_\alpha k_\beta\right)$$

$$= -\frac{1}{2} g^{\mu\alpha} g^{\nu\beta} H^2 R \left(\delta_{\mu 1}\delta_{\nu 0} - \delta_{\mu 0}\delta_{\nu 1}\right) H^2 R \left(\delta_{\alpha 1}\delta_{\beta 0} - \delta_{\alpha 0}\delta_{\beta 1}\right)$$

$$= -\frac{H^4 R^2}{2} g^{\mu\alpha} g^{\nu\beta} \left(\delta_{\alpha 0}\delta_{\beta 1}\delta_{\mu 0}\delta_{\nu 1} - \delta_{\alpha 0}\delta_{\beta 1}\delta_{\mu 1}\delta_{\nu 0} - \delta_{\alpha 1}\delta_{\beta 0}\delta_{\mu 0}\delta_{\nu 1}\right.$$

$$\left. + \delta_{\alpha 1}\delta_{\beta 0}\delta_{\mu 1}\delta_{\nu 0}\right)$$

$$= \frac{H^4 R^2}{2} \left[g^{00}g^{11} - 2(g^{01})^2 + g^{11}g^{00}\right] = H^4 R^2 .$$

On the Killing horizon $R = H^{-1}$, it is $\kappa_{Killing}^2 = H^2$ and Eq. (3.128) follows by choosing the positive sign arising from the square root of this equation.   □

3.12   Thermodynamics of Cosmological Horizons in General Relativity

One of the most interesting developments of black hole physics is black hole thermodynamics. Originally developed for stationary *event* horizons, this branch of theoretical physics is being extended to other types of horizons such as apparent, trapping, isolated, dynamical, and slowly evolving horizons. It was realized early on that also cosmological horizons have thermodynamics associated to them, beginning with the static event horizon of de Sitter space [41]. It is claimed in the literature that this cosmological thermodynamics extends also to FLRW *apparent* horizons.

3.13   Thermodynamics of de Sitter Space

The thermodynamics of de Sitter space was studied by Gibbons and Hawking [41] (see also [11, 25, 27, 62, 63, 72, 76] and the references therein). The de Sitter horizon is endowed with temperature and entropy, like the Schwarzschild event horizon, as was deduced using Euclidean field theory techniques [41]. Gibbons and Hawking computed the thermal bath seen by a timelike geodesic observer in de Sitter space carrying a (scalar) particle detector confined to a small tube around the observer's worldline [41]. The result is the *de Sitter horizon temperature* T_H given by

$$k_B T_H = \frac{\hbar}{c} \frac{H}{2\pi} = \frac{\hbar}{c} \frac{\kappa_{\text{Killing}}}{2\pi} \tag{3.129}$$

where we have restored the fundamental constants and κ_{Killing} is the surface gravity defined using the Killing vector $(\partial/\partial T)^a$ of de Sitter space. The entropy of the de Sitter horizon, commonly referred to as *Gibbons-Hawking entropy*, is

$$S_H = \frac{k_B c^3}{\hbar G} \frac{\mathscr{A}_H}{4} = \frac{k_B c^3}{\hbar G} \frac{\pi}{H^2} \tag{3.130}$$

or

$$S_H = \frac{\hbar c}{k_B G} \frac{1}{4\pi T_H^2}, \tag{3.131}$$

where $\mathscr{A}_H = 4\pi R_{\text{EH}}^2 = 4\pi H^{-2}$ is the area of the event horizon. It is common to write

$$S_H = k_B \frac{\mathscr{A}_H}{4 l_{\text{Pl}}^2}, \tag{3.132}$$

where $l_{Pl} = \sqrt{G\hbar/c^3}$ is the Planck length and l_{Pl}^2 is interpreted as a "quantum of area". There is radiation with a thermal spectrum near the horizon with characteristic wavelength $\sim H^{-1}$ and $T \leq 10^{-28}$ K today. The entropy is constant in time, in agrement with the fact that de Sitter space is static in the region $0 \leq R \leq R_H$.

There is no extremal horizon analogous to that of extremal black holes because the metric contains a single parameter.

The Kodama vector in de Sitter space is immediately given by Eq. (2.76) which, compared with the de Sitter line element in static coordinates (3.115), yields

$$K^a = \left(\frac{\partial}{\partial T} \right)^a , \tag{3.133}$$

i.e., the Kodama vector coincides with the timelike Killing vector (3.118) of de Sitter space. The surface gravity generated by this Killing-Kodama field is $\kappa = H = \sqrt{\Lambda/3}$ (Eq. (3.128)) and then the corresponding temperature is $T_H = \dfrac{\kappa}{2\pi}$ in geometrized units. The first law of thermodynamics is satisfied: taking the internal energy $U = M_{MSH} = \dfrac{4\pi}{3} R^3 \rho$, where M_{MSH} is the Misner-Sharp-Hernandez mass and $V_H = 4\pi R_{AH}^3/3$ as the volume bounded by the event horizon, one has

$$dU = d\left(\frac{4\pi}{3} R_H^3 \rho \right) = d\left(\frac{4\pi\rho}{3H^3} \right) = 0$$

because both H and ρ are constant, while

$$dV_H = d\left(\frac{4\pi}{3H^3} \right) = 0 ,$$

while S_H is constant; dS_H, dU, and dV_H all vanish making the first law of thermodynamics rather trivial.

3.14 Thermodynamics of Apparent/Trapping Horizons in FLRW Space

The thermodynamic formulae valid for the de Sitter event (and apparent) horizon are taken and adapted by many authors to the non-static apparent horizon of FLRW space, which does not coincide with the event horizon (the latter may not even exist). The apparent horizon is often argued to be a causal horizon associated with gravitational temperature, entropy and surface gravity in dynamical spacetimes ([5, 15, 22, 40, 46, 47, 67] and references therein) and, if these arguments hold, then they would apply also to cosmological horizons. That the thermodynamics is ill-defined for the event horizon of FLRW space (except for de Sitter space) was argued in [15, 26, 39, 40, 85]. The authors of [51, 86] attempted to compute the Hawking radiation of the FLRW apparent horizon. References [20, 62] rederived it using the Hamilton-Jacobi method [3, 68, 83] in the Parikh-Wilczek approach originally developed for black hole horizons [73]. In this context the particle emission rate in the WKB approximation is the tunneling probability for the classically forbidden trajectories from inside to outside the horizon,

$$\Gamma \sim \exp\left(-\frac{2\,\mathrm{Im}(I)}{\hbar}\right) \simeq \exp\left(-\frac{\hbar\omega}{k_B T}\right),\tag{3.134}$$

where I is the Euclideanized action with imaginary part $\mathrm{Im}(I)$, ω is the angular frequency of the radiated quanta (taken, for simplicity, to be those of a massless scalar field, which is the simplest field to perform Hawking effect calculations), and the Hawking temperature is read off the expression of the Boltzmann factor, $k_B T = \dfrac{\hbar^2 \omega}{2\,\mathrm{Im}(I)}$. The particle energy $\hbar\omega$ is defined in an invariant way as $\omega = -K^a \nabla_a I$, where K^a is the Kodama vector, and the action I satisfies the Hamilton-Jacobi equation

$$h^{ab} \nabla_a I \nabla_b I = 0 .\tag{3.135}$$

Although the definition of energy is coordinate-invariant, it depends on the choice of time, here defined as the Kodama time, and ω is a "Kodama (angular) frequency". It is not yet established beyond doubt, however, that this prescription gives a correct and consistent thermodynamics.

There has been a rather large literature on the thermodynamics of FLRW spaces (e.g., [2, 19, 30] and references therein). A review of the thermodynamical properties of the FLRW apparent horizon, as well as the computation of the Kodama vector, Kodama-Hayward surface gravity, and Hawking temperature in various coordinate systems are given in Ref. [30]. The *Kodama-Hayward temperature of the FLRW apparent horizon* is given by

$$
\begin{aligned}
k_B T &= -\left(\frac{\hbar}{c}\right) \frac{R_{\mathrm{AH}}\left(H^2 + \frac{\dot{H}}{2} + \frac{k}{2a^2}\right)}{2\pi} = \left(\frac{\hbar}{24\pi c}\right) R_{\mathrm{AH}} \mathscr{R} \\
&\doteq \left(\frac{\hbar G}{c}\right) \frac{R_{\mathrm{AH}}}{3} (3P - \rho) .
\end{aligned}\tag{3.136}
$$

For a spatially flat universe this expression reduces to

$$k_B T = \left(\frac{\hbar G}{c}\right) \frac{[H + \dot{H}/(2H)]}{2\pi} \doteq \left(\frac{\hbar G}{c}\right) \frac{(\rho - 3P)}{3H} .\tag{3.137}$$

This temperature depends on the choice of surface gravity κ since $T = \kappa/2\pi$ in geometrized units and, as we have seen, there are several inequivalent prescriptions for this quantity. The choice of κ giving the temperature reported here is the Kodama-Hayward choice of Eq. (2.87) [30]. In fact, the Kodama-Hayward surface gravity given by Eq. (2.87) is

$$\kappa_{\text{Kodama}} = -\frac{R_{\text{AH}}}{2}\left(2H^2 + \dot{H} + \frac{k}{a^2}\right) = -\frac{R_{\text{AH}}}{12}\mathscr{R}. \tag{3.138}$$

Proof. Using comoving coordinates and the decomposition of the metric $ds^2 = h_{ab}dx^a dx^b + R^2 d\Omega_{(2)}^2$ $(a, b = 1, 2)$, where $h_{ab} = \text{diag}\left(-1, \frac{a^2}{1 - kr^2}\right)$, Eq. (2.87) gives

$$\kappa_{\text{Kodama}} = \frac{\sqrt{1 - kr^2}}{2a}\partial_a\left[\frac{a}{\sqrt{1 - kr^2}}h^{ab}\partial_b\left(a(t)r\right)\right]$$

$$= \frac{\sqrt{1 - kr^2}}{2a}\partial_a\left[\frac{a}{\sqrt{1 - kr^2}}\left(-\dot{a}r\delta^{a0} + \frac{(1 - kr^2)}{a}\delta^{a1}\right)\right]$$

$$= -\frac{1}{2}\left[\left(\frac{\dot{a}^2}{a^2} + \frac{\ddot{a}}{a}\right)ar + \frac{kr}{a}\right] = -\frac{R}{2}\left(2H^2 + \dot{H} + \frac{k}{a^2}\right).$$

$\square$

The Kodama-Hayward dynamical surface gravity (3.138) vanishes if the scale factor has the special form $a(t) = \sqrt{\alpha t^2 + \beta t + \gamma}$, where α, β, and γ are constants.

On the apparent horizon, Eq. (3.138) can be written as [19]

$$\kappa_{\text{Kodama}} = \frac{1}{2HR_{\text{AH}}}\left(\frac{\dot{R}_{\text{AH}}}{R_{\text{AH}}} - 2H\right) \tag{3.139}$$

using Eqs. (3.106) and (3.138). In the usual prescription of stationary spacetime which assigns to a horizon of area $\mathscr{A}$ the entropy $\mathscr{A}/4$ (in geometrized units), the putative entropy of the FLRW apparent horizon would be

$$S_{\text{AH}} = \left(\frac{k_B c^3}{\hbar G}\right)\frac{\mathscr{A}_{\text{AH}}}{4} = \left(\frac{k_B c^3}{\hbar G}\right)\frac{\pi}{H^2 + k/a^2}, \tag{3.140}$$

where

$$\mathscr{A}_{\text{AH}} = 4\pi R_{\text{AH}}^2 = \frac{4\pi}{H^2 + k/a^2} \tag{3.141}$$

is the area of the event horizon. The Hamiltonian constraint (3.5) gives

$$S_{\text{AH}} \doteq \frac{3}{8\rho}, \qquad \dot{S}_{\text{AH}} \doteq \frac{9H}{8\rho^2}(P + \rho). \tag{3.142}$$

In an expanding universe the apparent horizon entropy increases if $P + \rho > 0$, stays constant if $P = -\rho$, and decreases if the weak energy condition is violated, $P < -\rho$.

The horizon temperature is positive only if the Ricci scalar is, which is equivalent to equations of state satisfying $P < \rho/3$ for a perfect fluid in Einstein theory [35]. This is the condition for the apparent horizon to be also a trapping horizon. A "cold horizon" with $T = 0$ is obtained for vanishing Ricci scalar but the entropy is positive for such an horizon. These properties induce caution in regarding Eqs. (3.136) and (3.140) as established.

What about the first law of thermodynamics for the apparent horizon of FLRW space? Several authors argue that the natural choice of surface gravity is the Kodama-Hayward one, which produces the apparent horizon temperature (3.136). The entropy usually attributed to the apparent horizon is $S_{\mathrm{AH}} = \dfrac{\mathscr{A}_{\mathrm{AH}}}{4} = \pi R_{\mathrm{AH}}^2$, and the internal energy U should be identified with the Misner-Sharp-Hernandez mass $M_{\mathrm{AH}} = \dfrac{4\pi R_{\mathrm{AH}}^3}{3} \rho$ contained inside the apparent horizon. As already remarked, the factor $\dfrac{4\pi R_{\mathrm{AH}}^3}{3}$ appearing in this espression is not the proper volume of a sphere of areal radius R_{AH} unless the universe has flat spatial sections, $k = 0$. This fact points us to use the areal volume

$$V_{\mathrm{AH}} \equiv \frac{4\pi R_{\mathrm{AH}}^3}{3} \tag{3.143}$$

instead of the proper volume in thermodynamics (failing to do so would jeopardize the possibility to write the first law). However, even with this *caveat*, the first law does *not* assume the form

$$T_{\mathrm{AH}} \dot{S}_{\mathrm{AH}} = \dot{M}_{\mathrm{AH}} + P \dot{V}_{\mathrm{AH}} \tag{3.144}$$

that one might expect.

Let us proceed to illustrate what the Kodama-Hayward proposal for T entails.

0th law of thermodynamics. The 0th law states that the temperature (or, equivalently, the surface gravity) is constant on the horizon. This law ensures that all points of the horizon are at the same temperature, or that there is no temperature gradient on it. This statement corresponds to thermal equilibrium on the horizon and is a rather trivial consequence of spherical symmetry.

1st law of thermodynamics. The first law of thermodynamics for apparent horizons is more complicated than (3.144) and was given in Refs. [46, 47] under the name of "unified first law". While using the Misner-Sharp-Hernandez mass M_{AH} enclosed by the apparent horizon as internal energy, the Kodama-Hayward horizon temperature (3.136), the areal volume (3.143), one introduces further quantities as follows. Decompose the metric as in Eq. (2.71); then the *work density* is

$$w_0 \equiv -\frac{1}{2} T_{ab} h^{ab} \tag{3.145}$$

(proportional to the trace of the matter stress-energy tensor over the 2-space normal
to the 2-spheres of symmetry);

$$\psi_a \equiv T_a{}^b \nabla_b R + w_0 \nabla_a R \tag{3.146}$$

is the energy flux (localized Bondi flux) across the apparent horizon, when computed
on this hypersurface. The quantity $\mathscr{A}_{\text{AH}} \psi_a$ is called the *energy supply vector*. If K^a
denotes the Kodama vector,

$$j_a \equiv \psi_a + w_0 K_a \tag{3.147}$$

is a divergence-free energy-momentum vector which can be used in lieu of ψ_a. The
Einstein equations then give [46, 47]

$$M_{\text{MSH}} = \kappa R^2 + 4\pi R^3 w_0 , \tag{3.148}$$

$$\nabla_a M_{\text{MSH}} = \mathscr{A} j_a . \tag{3.149}$$

The last equation is rewritten as [46, 47]

$$\mathscr{A} \psi_a = \nabla_a M_{\text{MSH}} - w_0 \nabla_a V_{\text{AH}} \tag{3.150}$$

("unified first law"). The energy supply vector is then written as

$$\mathscr{A} \psi_a = \frac{\kappa}{2\pi} \nabla_a \left(\frac{\mathscr{A}}{4} \right) + R \nabla_a \left(\frac{M_{\text{MSH}}}{R} \right) . \tag{3.151}$$

Along the apparent horizon, defined by $\nabla^c R \nabla_c R = 0$, it is

$$M_{\text{AH}} = \frac{R}{2} \left(1 - \nabla^c R \nabla_c R \right) |_{\text{AH}} = \frac{R_{\text{AH}}}{2}$$

and

$$\mathscr{A}_{\text{AH}} \psi_a = \frac{\kappa}{2\pi} \nabla_a S_{\text{AH}} = T_{\text{AH}} \nabla_a S_{\text{AH}} .$$

This equation is interpreted by saying that the energy supply across the apparent
horizon $\mathscr{A}_{\text{AH}} \psi_a$ is the "heat" $T_{\text{AH}} \nabla_a S_{\text{AH}}$ gained. Writing the energy supply explicitly
gives

$$T_{\text{AH}} \nabla_a S_{\text{AH}} = \nabla_a M_{\text{AH}} - w_0 \nabla_a V_{\text{AH}} \tag{3.152}$$

and $-w_0 \nabla_a V_{\text{AH}}$ is a work term. The "heat" entering the apparent horizon goes into
changing the internal energy M_{AH} and performing work due to the change in size of
this horizon.

Let us compute now the time component of Eq. (3.152) in comoving coordinates for a FLRW space sourced by a perfect fluid in General Relativity. We have $h_{ab} = \text{diag}\left(-1, \frac{a^2}{1-kr^2}\right)$ and

$$
\begin{aligned}
w_0 &\equiv -\frac{1}{2}\left[(P+\rho)u_a u_b + P g_{ab}\right]h^{ab} \\
&= -\frac{1}{2}\left[(P+\rho)h^{00}(u_0)^2 + P\left(h_{00}h^{00} + h_{11}h^{11}\right)\right] \\
&= -\frac{1}{2}\left[-(P+\rho) + 2P\right] \\
&= \frac{\rho - P}{2}.
\end{aligned}
\tag{3.153}
$$

One computes also

$$
\begin{aligned}
\dot{V}_{\text{AH}} &= 3V_{\text{AH}}\frac{\dot{R}_{\text{AH}}}{R_{\text{AH}}} = 3HV_{\text{AH}}\left(1 - \frac{\ddot{a}}{a}R_{\text{AH}}^2\right) \\
&= \frac{3HV_{\text{AH}}}{2\rho}\cdot 3(P+\rho)
\end{aligned}
\tag{3.154}
$$

and, using the last equation,

$$
\begin{aligned}
\dot{M}_{\text{AH}} &= \frac{d}{dt}(V_{\text{AH}}\rho) = \dot{V}_{\text{AH}}\rho + V_{\text{AH}}\dot{\rho} \\
&= \frac{3HV_{\text{AH}}}{2}(P+\rho).
\end{aligned}
\tag{3.155}
$$

We also have

$$
\dot{S}_{\text{AH}} = 2\pi R_{\text{AH}}\dot{R}_{\text{AH}} \doteq \frac{3\pi R_{\text{AH}}^2}{\rho}H(P+\rho)
\tag{3.156}
$$

and, using this relation,

$$
\begin{aligned}
T_{\text{AH}}\dot{S}_{\text{AH}} &= \frac{R_{\text{AH}}}{3}(3P-\rho)\,3\pi\frac{R_{\text{AH}}^2}{\rho}H(P+\rho) \\
&= \frac{3HV_{\text{AH}}}{4\rho}(P+\rho)(3P-\rho).
\end{aligned}
\tag{3.157}
$$

Therefore we have, using Eqs. (3.157), (3.155), and (3.154),

$$
T_{\text{AH}}\dot{S}_{\text{AH}} = \dot{M}_{\text{AH}} + \frac{(P-\rho)}{2}\dot{V}_{\text{AH}}.
\tag{3.158}
$$

In the infinitesimal interval of comoving time dt the changes in the thermodynamical quantities are related by

$$T_{\mathrm{AH}}dS_{\mathrm{AH}} = dM_{\mathrm{AH}} + dW_{\mathrm{AH}}, \qquad dW_{\mathrm{AH}} = \frac{(P - \rho)}{2} \, dV_{\mathrm{AH}}.$$

The coefficient of dV_{AH}, i.e., $-w_0 = (P - \rho)/2$ equals the pressure P (the naively expected coefficient) only if $P = -\rho$ (the case of de Sitter space in which dM_{AH}, dV_{AH}, and dS_{AH} all vanish). The fact that the coefficient appearing in the work term is not simply P can be understood as a consequence of the fact that the apparent horizon is not comoving [35]. For a comoving sphere of radius R_s it is $\dot{R}_s/R_s = H$ and $\dot{V}_s = 3HV_s$, while

$$\dot{M}_s = \frac{d}{dt}(V_s\rho) = \dot{V}_s\rho + V_s\dot{\rho} = 3HV_s\rho - 3HV_s(P + \rho) = -3HV_sP,$$

hence $\dot{M}_s + P\dot{V}_s = 0$. Indeed, the covariant conservation equation (3.7) is often presented as the first law of thermodynamics for a comoving volume V. Because of spatial homogeneity and isotropy there can be no preferred directions and physical spatial vectors in FLRW space, therefore the heat flux through a comoving volume must be zero. In fact, consider a comoving volume V_c (which, by definition, is constant in time) and the corresponding proper volume at time t, $V = a^3(t)V_c$. Multiplying Eq. (3.7) by V one obtains

$$a^3 V_c\dot{\rho} + 3a^2\dot{a}V_c(P + \rho) = 0$$

or

$$V\dot{\rho} + \dot{V}(P + \rho) = \frac{d}{dt}(\rho V) + P\dot{V} = 0.$$

By interpreting $U \equiv \rho V$ as the total internal energy of matter in the volume V one obtains the relation between variations in the time dt

$$dU + PdV = 0, \tag{3.159}$$

and the first law (with work term coefficient P) then gives $TdS = 0$, which is consistent with the above-mentioned absence of entropy flux vectors and with the well known fact that, in curved space, there is no entropy generation in a perfect fluid (the entropy along fluid lines remains constant and there is no exchange of entropy between neighbouring fluid lines [80]). Indeed, Eq. (3.62) for the evolution of the Misner-Sharp-Hernandez mass contained in a *comoving* sphere reduces to $\dot{M}_{\mathrm{MSH}} + P\dot{V} = 0$ or $\dot{\rho} + 3H(P + \rho) = 0$. However, for a non-comoving volume, the work term is more complicated than PdV.

Attempts to write the first law for the event, instead of the apparent, horizon lead to inconsistencies [15, 26, 39, 85]. This fact supports the belief that it is the apparent horizon which is the relevant quantity in the thermodynamics of cosmological horizons.

(Generalized) 2nd law of thermodynamics.

A second law of thermodynamics for the de Sitter horizon was already given in the original Gibbons-Hawking paper [41] and re-proposed in [65]. Davies [26] has considered the event horizon of FLRW space and, for General Relativity with a perfect fluid as the source, has proved the following theorem: if the cosmological fluid satisfies $P + \rho \geq 0$ and $a(t) \to +\infty$ as $t \to +\infty$, then the area of the event horizon is non-decreasing. The entropy of the event horizon is taken to be $S_{EH} = \left(\dfrac{k_B c^3}{\hbar G} \right) \dfrac{\mathscr{A}_{EH}}{4}$, where $\mathscr{A}_{EH}$ is the area of the event horizon. The validity of the generalized 2nd law for certain radiation-filled universes was discussed in [28, 29]. For radiation, the energy density is $\rho_{rad} = \dfrac{4\sigma}{c} T^4$, where σ is the Stefan-Boltzmann constant, and the entropy density[12] is

$$s_{rad} = \frac{4}{3} \frac{\rho_{rad}}{T} . \qquad (3.160)$$

The entropy of the radiation contained inside the volume V_{EH} enclosed by the event horizon is

$$S_{rad} = s_{rad} V_{EH} = \frac{4\sqrt{2}}{3} \left(\frac{\sigma}{c} \right)^{1/4} \rho_{rad}^{3/4} V_{EH} .$$

The total entropy of radiation contained in a comoving volume would stay constant because $\rho_{rad} \sim a^{-4}$ and $s_{rad} \sim a^3$ while a proper volume scales as $V \sim a^3$; however, the event horizon is not comoving and the radiation entropy within it decreases as V_{EH} expands slower than comoving (according to $\dot{R}_{EH}/R_{EH} = H - R_{EH}^{-1}$) and radiation crosses outside the event horizon. For realistic universes the entropy of the event horizon is much larger than the radiation entropy and, for universes departing slightly from a de Sitter universe due to the presence of a cosmological constant in addition to radiation, it can be shown analytically that the generalized 2nd law is valid, $\dot{S}_{rad} + \dot{S}_{EH} > 0$ [28, 29]. However, a general proof is not available.

Another question raised in [29] is the following: if the universe contains a gas of black holes, there is an entropy loss when these cross outside the horizon. For small black holes, this loss in more than compensated by the increase in area of the event horizon. However, for larger black holes, a preliminary study suggests a violation of the generalized 2nd law in certain open universes. These results cannot be fully relied upon because they assume relations valid for the Schwarzschild-de Sitter

---

[12]In general, the entropy density of a perfect fluid is $s = \dfrac{P + \rho}{T}$ [54].

black hole, which probably do not hold for large dynamical black holes embedded in a FLRW background. However, this issue remains unsolved and deserves further investigation with more reliable models.

Due to the difficulties with the event horizon one is led to consider the apparent horizon instead. Then, Eq. (3.156) tells us that, in an expanding universe in Einstein theory with perfect fluid, the apparent horizon area increases except for the quantum vacuum equation of state $P = -\rho$ (for which S_{AH} stays constant) and for phantom fluids with $P < -\rho$, in which case S_{AH} *decreases*, adding another element of weirdness to the behaviour of phantom matter.

The *generalized 2nd law* states that the total entropy of matter and of the horizon $S_{total} = S_{matter} + S_{AH}$ cannot decrease in any physical process,

$$\delta S = \delta S_{matter} + \delta S_{AH} \geq 0. \qquad (3.161)$$

(We refer here to the apparent horizon but several authors refer instead to the event or particle horizons. It is clear that the apparent horizon is more appropriate since it is a quantity defined quasi-locally and it always exists.) There is no definitive proof that the apparent horizon thermodynamics is consistent.

Thermodynamics of spacetime and cosmic holography. In the spirit of the thermodynamics of spacetime [34, 50], the Einstein-Friedmann equations of cosmology in Einstein theory have been derived from the first law of thermodynamics (3.158), first for spatially flat universes [15, 23] and then for general curvature index k [19]. An earlier work by Verlinde [82] derived the Einstein-Friedmann equations in a radiation-dominated FLRW universe from the Cardy-Verlinde formula, which gives the entropy for a conformal field theory, in the spirit of the holographic principle. The Fischler-Susskind version of this principle can be formulated by saying that the matter entropy contained in the volume enclosed by the particle horizon cannot exceed the entropy of the particle horizon itself. This principle restricts the matter content of the universe.

The cosmic holographic inequality is written as

$$s V_{PH} \leq \frac{\mathscr{A}_{PH}}{4}, \qquad (3.162)$$

where s is the entropy density of matter. Fischler and Susskind [37] find that the cosmic holographic principle is violated for fluids with $P = w\rho$ if $w < 1/3$. Bak and Rey instead apply the holographic inequality to the *apparent* horizon [5] with the following results. For $k = 0$ or -1, the cosmic holographic inequality $s V_{AH} \leq \mathscr{A}_{AH}/4$ is satisfied for perfect fluids with $|w| \leq 1$ (phantom fluids, in particular, violate the holographic bound) and if the inequality is satisfied at the Planck time [5]. Hence inflation, with $w \simeq -1$ violates also this version of the cosmic holographic principle.

Various entropy bounds have been discussed for FLRW space [5, 7, 13, 14, 17, 18, 33, 38, 52, 53, 58, 74, 81, 84]. These discussions use the particle horizon or the event horizon and seem to ignore the apparent horizon. Given the highly speculative nature of the subject and the fact that results in this area do not seem to be settled, we will not discuss it further and we refer the reader to these references.

3.15 Conclusions

The thermodynamics of apparent horizons is very intriguing, but is it correct? Is the entire construction consistent? After all, the work term appearing in the first law is somehow introduced a posteriori and, if one did not already know the first law, it is unlikely that one would discover it this way.

The apparent horizon thermodynamics is formulated for horizons changing in an arbitrary way. However, it would seem that equilibrium thermodynamics could only be introduced for physical systems in equilibrium and near equilibrium, and therefore the appropriate constructs should be slowly varying apparent horizons. In the literature, this restriction appears only in the Hamilton-Jacobi approach to the Wilczek-Parikh tunneling formalism, in which an high frequency approximation is used which requires that the background is varying slowly (although usually this requirement is not spelled out).

Apparent horizons seem to have implications also for the black hole information loss paradox and are seen as an alternative to firewalls [43], but this viewpoint need much work to be developed.

Problems

3.1. Compute the Misner-Sharp-Hernandez mass for a sphere in FLRW space using comoving coordinates.

3.2. In a $k = 0$ FLRW universe with a perfect fluid and constant equation of state $P = w\rho$ and $-1 < w < -1/3$ (accelerating but not superaccelerating universe), show that the event horizon is always outside the apparent horizon and is, therefore, unobservable.[13]

3.3. Compute the surface gravity of de Sitter space (3.128) using the property (2.78).

3.4. Check that Eq. (3.144) is not satisfied in FLRW space.

---

[13]Cf. Refs. [15, 85].

References

1. Akbar, M., Cai, R.-G.: Friedmann equations of FRW universe in scalar-tensor gravity, f(R) gravity and first law of thermodynamics. Phys. Lett. B **635**, 7 (2006)
2. Akbar, M., Cai, R.-G.: Thermodynamic behavior of the Friedmann equation at the apparent horizon of the FRW universe. Phys. Rev. D **75**, 084003 (2007)
3. Angheben, M., Nadalini, M., Vanzo, L., Zerbini, S.: Hawking radiation as tunneling for extremal and rotating black holes. J. High Energy Phys. **05**, 014 (2005)
4. Ashtekar, A., Krishnan, B.: Isolated and dynamical horizons and their applications. Living Rev. Relat. **7**, 10 (2004)
5. Bak, D., Rey, S.-J.: Cosmic holography. Class. Quantum Grav. **17**, L83 (2000)
6. Balasubramanian, V., de Boer, J., Minic, D.: Mass, entropy and holography in asymptotically de Sitter spaces. Phys. Rev. D **65**, 123508 (2002)
7. Banks, T., Fischler, W.: An upper bound on the number of e-foldings. Preprint arXiv:astro-ph/0307459
8. Ben-Dov, I.: Outer trapped surfaces in Vaidya spacetimes. Phys. Rev. D **75**, 064007 (2007)
9. Bengtsson, I., Senovilla, J.M.M.: Region with trapped surfaces in spherical symmetry, its core, and their boundaries. Phys. Rev. D **83**, 044012 (2011)
10. Bhattacharya, S., Lahiri, A.: On the existence of cosmological event horizons. Class. Quantum Grav. **27**, 165015 (2010)
11. Birrell, N.D., Davies, P.C.W.: Quantum Fields in Curved Space. Cambridge University Press, Cambridge (1982)
12. Booth, I., Brits, L., Gonzalez, J.A., Van den Broeck, V.: Black hole boundaries. Class. Quantum Grav. **23**, 413 (2006)
13. Bousso, R.: A covariant entropy conjecture. J. High Energy Phys. **07**, 004 (1999)
14. Bousso, R.: Positive vacuum energy and the N bound. J. High Energy Phys. **11**, 038 (2000)
15. Bousso, R.: Cosmology and the S-matrix. Phys. Rev. D **71**, 064024 (2005)
16. Bousso, R.: Adventures in de Sitter space. Preprint arXiv:hep-th/0205177
17. Brustein, R., Veneziano, G.: Causal entropy bound for a spacelike region. Phys. Rev. Lett. **84**, 5695 (2000)
18. Cai, R.-G.: Holography, the cosmological constant and the upper limit of the number of e-foldings. J. Cosmol. Astropart. Phys. **02**, 007 (2004)
19. Cai, R.-G., Kim, S.P.: First law of thermodynamics and Friedmann equations of Friedmann-Robertson-Walker universe. J. High Energy Phys. **0502**, 050 (2005)
20. Cai, R.-G., Cao, L.-M., Hu, Y.-P.: Hawking radiation of an apparent horizon in a FRW universe. Class. Quantum Grav. **26**, 155018 (2009)
21. Chakraborty, S., Mazumder, N., Biswas, R.: Cosmological evolution across phantom crossing and the nature of the horizon. Astrophys. Sp. Sci. **334**, 183 (2011)
22. Collins, W.: Mechanics of apparent horizons. Phys. Rev. D **45**, 495 (1992)
23. Danielsson, U.H.: Transplanckian energy production and slow roll inflatioon. Phys. Rev. D **71**, 023516 (2005)
24. Das, A., Chattopadhyay, S., Debnath, U.: Validity of generalized second law of thermodynamics in the logamediate and intermediate scenarios of the universe. Found. Phys. **42**, 266 (2011)
25. Davies, P.C.W.: Mining the universe. Phys. Rev. D **30**, 737 (1984)
26. Davies, P.C.W.: Cosmological horizons and entropy. Class. Quantum Grav. **5**, 1349 (1988)
27. Davies, P.C.W., Ford, L.H., Page, D.N.: Gravitational entropy: beyond the black hole. Phys. Rev. D **34**, 1700 (1986)
28. Davis, T.M., Davies, P.C.W.: How far can the generalized second law be generalized? Found. Phys. **32**, 1877 (2002)
29. Davis, T.M., Davies, P.C.W., Lineweaver, C.H.: Black hole versus cosmological horizon entropy. Class. Quantum Grav. **20**, 2753 (2003)
30. Di Criscienzo, R., Hayward, S.A., Nadalini, M., Vanzo, L., Zerbini, S.: On the Hawking radiation as tunneling for a class of dynamical black holes. Class. Quantum Grav. **27**, 015006 (2010)

31. d'Inverno, R.: Introducing Einstein's Relativity. Oxford University Press, Oxford (2002)
32. Doran, C.: A new form of the Kerr solution. Phys. Rev. D **61**, 06750 (2000)
33. Easther, R., Lowe, D.A.: Holography, cosmology, and the second law of thermodynamics. Phys. Rev. Lett. **82**, 4967 (1999)
34. Eling, C., Guedens, R., Jacobson, T.: Nonequilibrium thermodynamics of spacetime. Phys. Rev. Lett. **96**, 121301 (2006)
35. Faraoni, V.: Apparent and trapping cosmological horizons. Phys. Rev. D **84**, 024003 (2011)
36. Faraoni, V., Nielsen, A.B.: Quasi-local horizons, horizon-entropy, and conformal field redefinitions. Class. Quantum Grav. **28**, 175008 (2011)
37. Fischler, W., Susskind, L.: Holography and cosmology. Preprint arXiv:hep-th/9806039
38. Fischler, W., Loewy, A., Paban, S.: The entropy of the microwave background and the acceleration of the universe. J. High Energy Phys. **09**, 024 (2003)
39. Frolov, A., Kofman, L.: Inflation and de Sitter thermodynamics. J. Cosmol. Astropart. Phys. **0305**, 009 (2003)
40. Ghersi, J.T.G., Geshnizjani, G., Piazza, F., Shandera, S.: Eternal inflation and a thermodynamic treatment of Einstein's equations. J. Cosmol. Astropart. Phys. **1106**, 005 (2011)
41. Gibbons, G.W., Hawking, S.W.: Cosmological event horizon, thermodynamics, and particle creation. Phys. Rev. D **15**, 2738 (1977)
42. Guth, A.H.: The inflationary universe: a possible solution to the horizon and flatness problems. Phys. Rev. D **23**, 347 (1981)
43. Hawking, S.W.: Information preservation and weather forecasting for black holes. Preprint arXiv:1401.5761
44. Hawking, S.W., Ellis, G.F.R.: The Large Scale Structure of Space-Time. Cambridge University Press, Cambridge (1973)
45. Hayward, S.A.: Gravitational energy in spherical symmetry. Phys. Rev. D **53**, 1938 (1996)
46. Hayward, S.A.: Unified first law of black-hole dynamics and relativistic thermodynamics. Class. Quantum Grav. **15**, 3147 (1998)
47. Hayward, S.A., Mukohyama, S., Ashworth, M.C.: Dynamic black holes and entropy. Phys. Lett. A **256**, 347 (1999)
48. Ibison, M.: On the conformal forms of the Robertson-Walker metric. J. Math. Phys. **48**, 122501 (2007)
49. Infeld, L., Schild, A.: A new approach to kinematic cosmology. Phys. Rev. **68**, 250 (1945)
50. Jacobson, T.: Thermodynamics of spacetime: the Einstein equation of state. Phys. Rev. Lett. **75**, 1260 (1995)
51. Jang, K.-X., Feng, T., Peng, D.-T.: Hawking radiation of apparent horizon in a FRW universe as tunneling beyond semiclassical approximation. Int. J. Theor. Phys. **48**, 2112 (2009)
52. Kaloper, N., Linde, A.D.: Cosmology versus holography. Phys. Rev. D **60**, 103509 (1999)
53. Kaloper, N., Kleban, M., Sorbo, L.: Observational implications of cosmological event horizons. Phys. Lett. B **600**, 7 (2004)
54. Kolb, E.W., Turner, M.S.: The Early Universe. Addison-Wesley, Reading (1990)
55. Kraus, P., Wilczek, F.: Some applications of a simple stationary line element for the Schwarzschild geometry. Mod. Phys. Lett. A **9**, 3713 (1995)
56. Li, R., Ren, J.-R., Shi, D.-F.: Fermions tunneling from apparent horizon of FRW universe. Phys. Lett. B **670**, 446 (2009)
57. Liddle, A.R., Lyth, D.H.: Cosmological Inflation and Large Scale Structure. Cambridge University Press, Cambridge (2000)
58. Lowe, D.A., Marolf, D.: Holography and eternal inflation. Phys. Rev. D **70**, 026001 (2004)
59. Martel, K., Poisson, E.: Regular coordinate systems for Schwarzschild and other spherical spacetimes. Am. J. Phys. **69**, 476 (2001)
60. Mazumder, N., Biswas, R., Chakraborty, S.: Interacting three fluid system and thermodynamics of the universe bounded by the event horizon. Gen. Rel. Gravit. **43**, 1337 (2011)
61. McVittie, G.C.: The mass-particle in an expanding universe. Mon. Not. R. Astr. Soc. **93**, 325 (1933)

62. Medved, A.J.M.: Radiation via tunneling from a de Sitter cosmological horizon. Phys. Rev. D **66**, 124009 (2002)
63. Medved, A.J.M.: A brief editorial on de Sitter radiation via tunneling. Preprint arXiv:0802.3796
64. Mosheni Sadjadi, H.: Generalized second law in a phantom-dominated universe. Phys. Rev. D **73**, 0635325 (2006)
65. Mottola, E.: Thermodynamic instability of de Sitter space. Phys. Rev. D **33**, 1616 (1986)
66. Mukhanov, V.: Physical Foundations of Cosmology. Cambridge University Press, Cambridge (2005)
67. Nielsen, A.B., Yeom, D.-H.: Black holes without boundaries. Int. J. Mod. Phys. A **24**, 5261 (2009)
68. Nielsen, A.B., Visser, M.: Production and decay of evolving horizons. Class. Quantum Grav. **23**, 4637 (2006)
69. Nolan, B.C.: Sources for McVittie's mass particle in an expanding universe. Phys. Rev. D **58**, 064006 (1998)
70. Nolan, B.C.: A Point mass in an isotropic universe. 2. Global properties. Class. Quantum Grav. **16**, 1227 (1999)
71. Nolan, B.C.: A Point mass in an isotropic universe 3. The region $R \leq 2m$. Class. Quantum Grav. **16**, 3183 (1999)
72. Parikh, M.K.: New coordinates for de Sitter space and de Sitter radiation. Phys. Lett. B **546**, 189 (2002)
73. Parikh, M.K., Wilczek, F.: Hawking radiation as tunneling. Phys. Rev. Lett. **85**, 5042 (2000)
74. Piao, Y.-S.: Entropy of the microwave background radiation in the observable universe. Phys. Rev. D **74**, 47301 (2006)
75. Rindler, W.: Visual horizons in world-models. Mon. Not. R. Astr. Soc. **116**, 663 (1956) (reprinted in Gen. Rel. Gravit. **34**, 133 (2002))
76. Sekiwa, Y.: Decay of the cosmological constant by Hawking radiation as quantum tunneling. Preprint arXiv:0802.3266
77. Senovilla, J.M.M.: Singularity theorems and their consequences. Gen. Rel. Grav. **30**, 701 (1998)
78. Spradlin, M.A., Strominger, A., Volovich, A.: Les Houches lectures on de Sitter space. In Les Houches 2001, Gravity, Gauge Theories and Strings, Proceedings of the Les Houches Summer School 76, Les Houches, pp. 423–453 (2001). (Preprint arXiv:hep-th/011007)
79. Spradlin, M.A., Volovich, A.: Vacuum states and the S matrix in dS/CFT. Phys. Rev. D **65**, 104037 (2002)
80. Stephani, H.: General Relativity. Cambridge University Press, Cambridge (1982)
81. Veneziano, G.: Pre-bangian origin of our entropy and time arrow. Phys. Lett. B **454**, 22 (1999)
82. Verlinde, E.: On the holographic principle in a radiation-dominated universe. Preprint arXiv:hep-th/0008140
83. Visser, M.: Essential and inessential features of Hawking radiation. Int. J. Mod. Phys. D **12**, 649 (2003)
84. Wang, B., Abdalla, E.: Plausible upper limit on the number of e-foldings. Phys. Rev. D **69**, 104014 (2004)
85. Wang, B., Gong, Y., Abdalla, E.: Thermodynamics of an accelerated expanding universe. Phys. Rev. D **74**, 083520 (2006)
86. Zhu, T., Ren, J.-R.: Corrections to Hawking-like radiation for a Friedmann-Robertson-Walker universe. Eur. Phys. J. C **62**, 413 (2009)

Chapter 4
Inhomogeneities in Cosmological "Backgrounds" in Einstein Theory

> *One has no right to love or hate anything if one has not acquired a thorough knowledge of its nature. Great love springs from great knowledge of the beloved object, and if you know it but little you will be able to love it only a little or not at all.*
>
> —Leonardo da Vinci

4.1 Introduction

There is much motivation for studying analytic solutions of General Relativity and of alternative theories of gravity representing central inhomogeneities embedded in cosmological "backgrounds". Our main interest is understanding apparent and trapping horizons and their dynamics. Another motivation comes from the fact that the present acceleration of the cosmic expansion [8, 87, 136, 137, 142–145, 162] requires, in the context of General Relativity, that approximately 73 % of the energy content of the universe is in the form of a mysterious dark energy [90] (see Ref. [5] for a discussion). Dark energy appears as an ad hoc explanation and an alternative to it could be that gravity deviates from General Relativity at large scales, which leads one to take more seriously alternative theories of gravity. Further motivation for alternative gravity comes from the fact that virtually all theories attempting to quantize gravity produce, in their low-energy limits, actions containing corrections to Einstein theory such as nonminimally coupled dilatons and/or higher derivative terms. These ideas have led to the introduction of $f(\mathscr{R})$ gravity in cosmology to modify Einstein theory at large scales [26, 30, 152–154, 156, 157, 168] and explain the cosmic acceleration without dark energy (see Refs. [40, 155] for reviews). Given that the $f(\mathscr{R})$ theories of interest for cosmology (which are the most relevant in today's theoretical physics landscape) are designed to produce a time-varying effective cosmological "constant", spherically symmetric solutions representing black holes or other inhomogeneities in these theories are expected to be dynamical and to have FLRW asymptotics, not to be static and asymptotically flat. Relatively speaking, very few such solutions are known. However, analytic solutions describing central objects in cosmological "backgrounds" are interesting also in General Relativity and not only in alternative gravity. The first study of this kind of solution by McVittie in 1933 [119] was motivated by the need to understand whether, and to

© Springer International Publishing Switzerland 2015
V. Faraoni, *Cosmological and Black Hole Apparent Horizons*,
Lecture Notes in Physics 907, DOI 10.1007/978-3-319-19240-6_4

what extent, the cosmic expansion affects the dynamics of local systems (Ref. [28] reviews this subject). This question constitutes independent motivation for studying black holes embedded in FLRW cosmologies. If objects which are strongly bound gravitationally (and nothing is more strongly bound than a black hole) become comoving and follow the expansion of the cosmic substratum, any weakly bound object should do the same. This issue can only be studied by analytic solutions of the Einstein equations if the central object is to be strongly bound.

The old McVittie solution [119] has been largely overlooked for decades and recent studies show that its structure and details are not yet completely understood [38, 55, 85, 94, 123, 124]. Relatively few other solutions describing central condensations in otherwise spatially homogeneous universes have been reported over the years (see Ref. [92] for an in-depth discussion of inhomogeneous cosmologies from a more general point of view).

Cosmological condensations in General Relativity have received recent attention also for other reasons, following other attempts to explain the present acceleration of the universe without exotic dark energy and without modifying gravity. The first idea consists of using the backreaction of inhomogeneities on the cosmic dynamics to produce the observed acceleration [20–24, 88, 97, 98, 101, 102, 134, 139, 140, 173, 174]. This backreaction idea is implemented in a formalism plagued by formal problems and it has not been demonstrated that it is able to explain the cosmic acceleration. The sign of the backreaction terms (let alone their magnitude) in the equation giving the averaged acceleration has not been shown to be the correct one [17, 96, 163, 167]. Even more serious doubts are cast on this proposed solution to the cosmic acceleration problem by the (admittingly more formal) work of Ref. [71].

A second idea to move beyond these riddles attributes the cosmic acceleration to the possibility that we live inside a giant void, which involves the consideration of analytic solutions of Einstein theory describing cosmological inhomogeneities (Lemaître-Tolman-Bondi, Swiss-cheese, and other models) [89, 92, 112–114, 133, 135, 139, 140]. There has also been interest in evolving horizons in relation to the accretion of dark energy [6, 32, 41, 69, 72, 78, 83, 108, 159]. Accretion onto a primordial black hole in the early universe could have been so rapid to make its growth very fast [27, 74–76]. The accretion of dark energy and, in particular, of phantom energy (if this extreme form of dark energy exists at all) by a black hole, and the consequent backreaction and mass change have been the subject of much recent literature. Analytic solutions which accrete from their surroundings are useful to elucidate questions in this area [66, 121].

Another major motivation for studying cosmological black holes is that explicit examples of time-varying horizons would be very useful to understand Hawking radiation and formulate black hole and horizon thermodynamics in fully dynamical situations.

Due to the non-linearity of the Einstein equations, it is impossible to split a metric into a cosmological "background" and a part describing a spherical inhomogeneity. However, in the solutions described in this chapter, the spacetime reduces to a FLRW universe when a parameter (related to the mass of the central inhomogeneity) vanishes and, therefore, we will loosely use the word "background" to denote this FLRW spacetime.

Let us proceed to analyze general-relativistic spacetimes which describe cosmological black holes and have time-evolving horizons at least part of the time. All these solutions of the Einstein equations are spherically symmetric.

4.2 Schwarzschild-de Sitter-Kottler Spacetime

The *line element of the Schwarzschild-de Sitter-Kottler spacetime* [91] is

$$ds^2 = -\left(1 - \frac{2m}{R} - H^2 R^2\right) dt^2 + \left(1 - \frac{2m}{R} - H^2 R^2\right)^{-1} dR^2 + R^2 d\Omega^2_{(2)}, \quad (4.1)$$

where R is obviously an areal radius, the constant $H = \sqrt{\Lambda/3}$ is the Hubble parameter of the de Sitter "background". $\Lambda > 0$ is the cosmological constant, and the positive parameter m describes the mass of the central inhomogeneity (e.g., [14, 77]). The static coordinates (t, R, θ, φ) cover the region

$$0 < t < +\infty,$$
$$R_1 < R < R_2,$$
$$0 < \theta < \pi,$$
$$0 < \varphi < 2\pi,$$

where $R_{1,2}$ denote the horizon radii, see below.

The usual recipe $g^{RR} = 0$ locates the apparent horizons, with radii given by the positive roots of the cubic

$$H^2 R^3 - R + 2m = 0. \quad (4.2)$$

The solutions are

$$R_1 = \frac{2}{\sqrt{3}H} \sin \psi, \quad (4.3)$$

$$R_2 = \frac{1}{H} \cos \psi - \frac{1}{\sqrt{3}H} \sin \psi, \quad (4.4)$$

$$R_3 = -\frac{1}{H} \cos \psi - \frac{1}{\sqrt{3}H} \sin \psi, \quad (4.5)$$

where $\sin(3\psi) = 3\sqrt{3}\, mH$. In an expanding universe, both m and H are positive, which implies that the root R_3 is negative and unphysical, hence there can be at most two apparent horizons. When R_1 and R_2 are real, R_1 is a black hole apparent horizon (it reduces to the Schwarzschild horizon $R = 2m$ in the limit $H \to 0$), while

R_2 is a cosmological apparent horizon (it becomes the static de Sitter horizon with $R = H^{-1}$ in the limit $m \to 0$). The Schwarzschild-de Sitter-Kottler line element is static in the region covered by the coordinates (t, R, θ, φ) and comprised between the black hole and the cosmological horizons.

Both apparent horizons exist only if $0 < \sin(3\psi) < 1$. Then, since the metric is static between the horizons, the black hole and cosmological apparent horizons are also event horizons, hence they are null surfaces. When $\sin(3\psi) = 1$ the two horizons coincide: this extremal case corresponds to the Nariai black hole [125, 126]. When instead $\sin(3\psi) > 1$, the radii of both apparent horizons assume complex values and are unphysical: the spacetime then contains a naked singularity. To summarize:

- If $mH < 1/(3\sqrt{3})$ there are two horizons with radii R_1 and R_2.
- If $mH = 1/(3\sqrt{3})$ it is $R_1 = R_2$ and the two horizons coincide.
- If $mH > 1/(3\sqrt{3})$ there are no apparent horizons.

The last situation in this list is interpreted by noting that the black hole horizon becomes larger than the cosmological horizon and any observer in the region $R_1 < R < R_2$ cannot know about it. The spacetime region below the cosmological horizon can only accommodate a black hole smaller than (or as large as) this horizon.

The cosmological apparent horizon has a smaller radius than the radius that the de Sitter apparent horizon would have were the black hole absent: $R_2 < H^{-1}$. What is more, the black hole apparent horizon is larger than that of a Schwarzschild black hole of the same mass m: $R_1 > 2m$. The physical interpretation is that the cosmological "background" stretches the horizon of a black hole embedded in it, while the black hole contracts the cosmological horizon.

The area of the black hole horizon $\mathscr{A} = 4\pi R_1^2$ is, of course, time-independent. The central singularity is eternal and the black hole event horizon (when it exists) surrounds it.

The Misner-Sharp-Hernandez mass (2.92) of a sphere of radius R is calculated as

$$M_{\text{MSH}} = m + \frac{H^2 R^3}{2} = m + \frac{4\pi}{3} \rho R^3 , \qquad (4.6)$$

where $\rho = \dfrac{\Lambda}{8\pi}$ is the energy density of the de Sitter "background".

The Schwarzschild-de Sitter-Kottler black hole is the subject of an extensive literature devoted to the thermodynamical properties of its horizons. The thermodynamics is particularly interesting because of the simultaneous presence of a black hole and a cosmological horizon (e.g., [151]). There are two distinct temperatures associated with the two horizons and it appears that thermal equilibrium is only possible for an extremal (Nariai) black hole, in which these two temperatures become equal. Black holes embedded in anti-de Sitter space have also been the subject of much recent interest due to the AdS/CFT correspondence and to the broader fluid-gravity duality (e.g., [59, 81]), but they will not be discussed here.

4.3 McVittie Solution

The McVittie spacetime discovered in 1933 [119] generalizes the Schwarzschild-de Sitter-Kottler solution and is commonly interpreted as describing a central object embedded in an FLRW space which, in general, is not de Sitter. Therefore, the geometry in the region between the black hole and the cosmological horizon is time-dependent. It was shown long ago that the McVittie spacetime is the only perfect fluid solution of the Einstein equations which is spherically symmetric, shear free, and asymptotically FLRW [141]. In spite of much work [1, 3, 38, 85, 92, 94, 123, 124, 130–132, 161], the McVittie spacetime still eludes our full understanding. A simplifying assumption stated explicitly by McVittie in constructing his solution is the no-accretion condition $G_0^1 = 0$ (in spherical coordinates). According to the Einstein equations this condition, equivalent to $T_0^1 = 0$, excludes the radial flow of cosmic fluid (although such flow would realistically occur when a spherical local overdensity alters the "background"). Generalizations of the McVittie solution allowing for the possibility of radial flow of energy are discussed later.

McVittie [119] intended to elucidate the extent of the effects of the cosmological expansion on local gravitationally bound systems. Other approaches to this problem generated various solutions of the Einstein equations, such as the Swiss-cheese model of Einstein and Straus [44, 45]. The issue of cosmological expansion versus local dynamics has been debated extensively but is not completely solved (see the review in Ref. [28]). A complication (and an opportunity for us) is that, unlike the Schwarzschild-de Sitter-Kottler spacetime, black holes in general FLRW "backgrounds" are dynamical.

The McVittie line element is

$$ds^2 = -\frac{\left(1 - \frac{m(t)}{2\bar{r}}\right)^2}{\left(1 + \frac{m(t)}{2\bar{r}}\right)^2}\, dt^2 + a^2(t)\left(1 + \frac{m(t)}{2\bar{r}}\right)^4 \left(d\bar{r}^2 + \bar{r}^2 d\Omega_{(2)}^2\right) \tag{4.7}$$

in isotropic coordinates, where the McVittie no-accretion condition[1] $G_0^1 = 0$ [119] dictates that the function $m(t)$ satisfy

$$\frac{\dot{m}}{m} + \frac{\dot{a}}{a} = 0, \tag{4.8}$$

hence it is

$$m(t) = \frac{m_0}{a(t)}, \tag{4.9}$$

---

[1] This is hypothesis *e)* of Ref. [119].

with m_0 a non-negative constant. The physical mass of the central object is the Misner-Sharp-Hernandez mass, while $m(t)$ is just a metric coefficient in the particular coordinates adopted. We can then write the *McVittie line element* in isotropic coordinates as

$$ds^2 = -\frac{\left[1 - \frac{m_0}{2\bar{r}a(t)}\right]^2}{\left[1 + \frac{m_0}{2\bar{r}a(t)}\right]^2} dt^2 + a^2(t)\left[1 + \frac{m_0}{2\bar{r}a(t)}\right]^4 \left(d\bar{r}^2 + \bar{r}^2 d\Omega^2_{(2)}\right). \qquad (4.10)$$

The McVittie spacetime becomes the Schwarzschild one written in isotropic coordinates if $a \equiv 1$, and it reduces to the FLRW metric when m vanishes. Apart from the special case of a de Sitter "background", the line element (4.7) is singular on the 2-sphere $\bar{r} = m/2$ (which reduces to the Schwarzschild horizon if $a \equiv 1$) [58, 130–132, 161]. This singularity is spacelike [130–132] and there is another spacetime singularity at $\bar{r} = 0$. McVittie originally interpreted the metric (4.7) as describing a point mass at $\bar{r} = 0$ but, in general, this point mass would be surrounded by the $\bar{r} = m/2$ singularity, the interpretation of which is elusive [58, 130–132, 161].

According to Nolan [130], this singularity is weak in the sense that an object falling across $\bar{r} = m/2$ is not crushed to zero volume and, therefore, the energy density of the surrounding fluid must be finite. However, it is undeniable that the pressure of this fluid

$$P = -\frac{1}{8\pi}\left[3H^2 + \frac{2\dot{H}\left(1 + \frac{m}{2\bar{r}}\right)}{1 - \frac{m}{2\bar{r}}}\right] \qquad (4.11)$$

diverges at $\bar{r} = m/2$ together with the Ricci scalar $\mathscr{R} = 8\pi\left(\rho - 3P\right)$ [58, 117, 118, 130–132, 161], violating the weak and null energy conditions (but not the positivity of the energy density) in a neighbourhood of the singularity. The de Sitter "background" is an exception: in this case it is $\dot{H} = 0$ identically and the second term on the right hand side of Eq. (4.11), which causes P to diverge, is absent [51, 130–132].

Let us rewrite the McVittie line element (4.7) in terms of the areal radius

$$R \equiv a(t)\bar{r}\left(1 + \frac{m}{2\bar{r}}\right)^2; \qquad (4.12)$$

the differentials $d\bar{r}$ and dR satisfy the relation

$$dR = \left(1 + \frac{m}{2\bar{r}}\right)a\bar{r}\left[H\left(1 + \frac{m}{2\bar{r}}\right) + \frac{\dot{m}}{\bar{r}}\right]dt + a\left(1 + \frac{m}{2\bar{r}}\right)\left(1 - \frac{m}{2\bar{r}}\right)dr.$$

Now Eq. (4.8) yields

$$H\left(1 + \frac{m}{2\bar{r}}\right) + \frac{\dot{m}}{\bar{r}} = H\left(1 - \frac{m}{2\bar{r}}\right) \qquad (4.13)$$

and

$$dr = \frac{dR}{a\left(1 + \frac{m}{2r}\right)\left(1 - \frac{m}{2r}\right)} - H\bar{r}dt .$$

(4.14)

Upon use of the identity

$$\left(\frac{1 - \frac{m}{2r}}{1 + \frac{m}{2r}}\right)^2 = 1 - \frac{2m_0}{R} ,$$

(4.15)

substitution into Eq. (4.7) and manipulations yields[2]

$$ds^2 = -\left(1 - \frac{2m_0}{R} - H^2R^2\right)dt^2 + \frac{dR^2}{1 - \frac{2m_0}{R}} - \frac{2HR}{\sqrt{1 - \frac{2m_0}{R}}}dtdR + R^2d\Omega_{(2)}^2 ,$$

(4.16)

where $H \equiv \dot{a}/a$ is the Hubble parameter of the FLRW "background". In order to remove the cross-term in $dtdR$ from the line element we define a new time coordinate $T(t, R)$ with the differential relation

$$dT = \frac{1}{F}\left(dt + \beta dR\right) ,$$

(4.17)

where $F(t, R)$ is an integrating factor and the function $\beta(t, R)$ must be determined so that in the new coordinates the time-radius component of the metric vanishes. dT is an exact differential if the 1-form (4.17) is closed,

$$\frac{\partial F}{\partial R} = -F\frac{\partial \beta}{\partial t} + \beta\frac{\partial F}{\partial t} .$$

(4.18)

The substitution $dt = FdT - \beta dR$ in Eq. (4.16) yields

$$ds^2 = -\left(1 - \frac{2M}{R} - H^2R^2\right)F^2dT^2$$

$$+ \left[-\left(1 - \frac{2M}{R} - H^2R^2\right)\beta^2 + \frac{1}{1 - \frac{2M}{R}} + \frac{2\beta HR}{\sqrt{1 - \frac{2M}{R}}}\right]dR^2$$

$$+ 2F\left[\left(1 - \frac{2M}{R} - H^2R^2\right)\beta - \frac{HR}{\sqrt{1 - \frac{2M}{R}}}\right]dTdR + R^2d\Omega_{(2)}^2 . \quad (4.19)$$

---

[2] Use $m/\bar{r} = ma/R = m_0/R$, where ma is constant.

We now choose

$$\beta(t, R) = \frac{HR}{\sqrt{1 - \frac{2m_0}{R} \left(1 - \frac{2m_0}{R} - H^2 R^2\right)}}, \qquad (4.20)$$

and the line element becomes

$$ds^2 = -\left(1 - \frac{2m_0}{R} - H^2 R^2\right) F^2 dT^2 + \frac{dR^2}{1 - \frac{2m_0}{R} - H^2 R^2} + R^2 d\Omega^2_{(2)}. \qquad (4.21)$$

For a de Sitter "background" with $H = $ constant, the integrating factor F can be set to unity, recovering the Schwarzschild-de Sitter-Kottler metric (4.1): clearly the latter is a special case of the McVittie metric (4.21). In the new coordinates, the spacetime singularity $\bar{r} = m/2$ corresponds to $R = 2m\,a(t) = 2m_0$ and does not expand with the cosmic substratum.

Using Eqs. (4.21) and (2.92), the Misner-Sharp-Hernandez mass of a sphere of symmetry with proper radius R is found to be

$$M_{\text{MSH}} = m_0 + \frac{H^2 R^3}{2} = m_0 + \frac{4\pi}{3} \rho R^3, \qquad (4.22)$$

which is interpreted as a time-independent contribution m_0 from the central object plus the mass of the cosmic fluid contained in the sphere. Except for the static de Sitter case, the Misner-Sharp-Hernandez mass of the sphere changes due to the fact that the contribution from the cosmic fluid changes (because the radius R changes in time, or because the density of the fluid itself varies with time, or both). In particular, apparent horizons are not comoving and the mass enclosed by a black hole apparent horizon (when this exists) changes even in the absence of accretion because of the time evolution of both the areal radius R_{AH} of the apparent horizon and of the time evolution of the density $\rho(t)$.

4.3.1 Apparent Horizons

We now restrict ourselves to a spatially flat FLRW "background" for simplicity and we assume a perfect fluid stress-energy tensor asymptotically. The Einstein equations then provide the energy density ρ and the pressure P of the McVittie fluid source. The density is the same as for the FLRW "background",

$$\rho(t) = \frac{3}{8\pi} H^2(t). \qquad (4.23)$$

The "background" perfect fluid can be assigned any equation of state. For the sake of illustration, however, we consider only a cosmic fluid which reduces to a timelike dust at spatial infinity, with equation of state parameter $w = 0$. The pressure is then [123, 124]

$$P(t, R) = \rho(t) \left(\frac{1}{\sqrt{1 - \frac{2m_0}{R}}} - 1 \right).$$ (4.24)

The $g^{RR} = 0$ recipe locates the apparent horizons of the McVittie metric and gives

$$H^2(t)\, R^3 - R + 2m_0 = 0.$$ (4.25)

This cubic equation is nothing but the Schwarzschild-de Sitter-Kottler horizon condition (4.2) but now with a time-dependent Hubble parameter $H(t)$. The radii of the time-dependent apparent horizons are labelled $R_1(t)$ and $R_2(t)$ and correspond to the solutions $R_{1,2}$ of Eq. (4.2) with the replacement $H \to H(t)$. Therefore, the location of the apparent horizons of the McVittie spacetime varies with the cosmic time.

As for the Schwarzschild-de Sitter-Kottler case, both horizons exist if $0 < \sin(3\psi) < 1$, which corresponds to $m_0 H(t) < 1/(3\sqrt{3})$. However, contrary to the Schwarschild-de Sitter-Kottler space with constant Hubble parameter, this inequality is only satisfied at certain times, but not at other times, during the cosmic history. There is a unique instant $t_* = 2\sqrt{3}\, m_0$ at which $m_0 H(t) = 1/(3\sqrt{3})$ in a dust-dominated "background" with $H(t) = 2/(3t)$. Three possibilities occur:

- Early on, as $t < t_*$, it is $m_0 > \dfrac{1}{3\sqrt{3}\, H(t)}$ and both $R_1(t)$ and $R_2(t)$ are complex. There are no apparent horizons.

- At the time t_* it is $m_0 = \dfrac{1}{3\sqrt{3}\, H(t)}$ and two apparent horizons $R_1(t)$ and $R_2(t)$ coincide at the real physical location $R_1 = R_2 = \dfrac{1}{\sqrt{3}\, H(t)}$.

- At late times $t > t_*$, we have $m_0 < \dfrac{1}{3\sqrt{3}\, H(t)}$ and both $R_1(t)$ and $R_2(t)$ are real. There are then two distinct apparent horizons.

Figure 4.1 illustrates the qualitative dynamics of the McVittie apparent horizons.

As there are no apparent horizons at early times $t < t_*$, a naked singularity raises its head at $R = 2m_0$, where the Ricci scalar and the pressure diverge. In fact, the Hubble parameter $H(t)$ diverges in the early universe and the mass m_0 stays supercritical, with $m_0 > \dfrac{1}{3\sqrt{3}\, H(t)}$. It seems appropriate to interpret this naked singularity on the lines of the one in the Schwarzschild-de Sitter-Kottler spacetime: a black hole horizon cannot be accommodated in a "universe" which is too small to contain it. At these early times the interpretation of the "background" as a "universe"

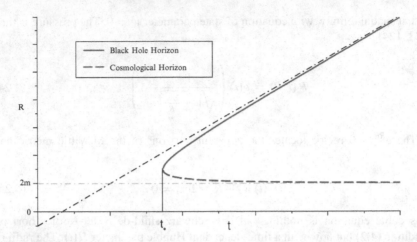

Fig. 4.1 The radii of the McVittie black hole (*continuous*) and cosmological (*dotted*) apparent horizons versus comoving time in a dust-dominated FLRW "background". Time t and radius R are in units of m and we arbitrarily fix $m = 1$

is not granted because the McVittie spacetime represents neither an isolated object nor a FLRW universe; at early times it is an inhomogeneous spacetime which is drastically different from both.

At the critical time t_* a black hole apparent horizon appears simultaneously with, and coinciding with, a cosmological apparent horizon

$$R_1(t_*) = R_2(t_*) = \frac{1}{\sqrt{3}\,H(t_*)}\,. \tag{4.26}$$

In the dust-dominated cosmological "background" it is easy to compute the radius of this horizon as $R_1 = R_2 = 3m_0$. This extremal situation resembles the Nariai black hole of the Schwarzschild-de Sitter-Kottler case, but it is instantaneous.

Later on, for $t > t_*$, the extremal horizon bifurcates into a black hole apparent horizon surrounded by a cosmological horizon (both evolving in time). The black hole apparent horizon contracts and asymptotes to the spacetime singularity $R = 2m_0$ from above as $t \to +\infty$. The cosmological apparent horizon expands monotonically, its radius approaching $1/H(t)$.

The singularity $R = 2m_0$ has been discussed extensively [94, 123, 124, 130–132]. The surface of equation $f(R) \equiv R - 2m_0 = 0$ has normal vector $N_\mu = \nabla_\mu f = \delta_{1\mu}$, which has norm squared

$$N_a N^a = g^{ab} N_a N_b \,|_{R=2m_0} = -4m_0^2 H^2(t) < 0\,. \tag{4.27}$$

As N^c is timelike, the singularity $R = 2m_0$ is spacelike. The curvature scalar

$$\mathscr{R} = -8\pi T_a^a = 8\pi \, (\rho - 3P) = 8\pi\rho(t) \left(4 - \frac{3}{\sqrt{1 - \frac{2m_0}{R}}} \right) \qquad (4.28)$$

diverges as $R \to 2m_0^+$. The two spacetime regions $R < 2m_0$ and $R > 2m_0$ are disconnected and separated by the $R = 2m_0$ singularity [130–132], with the geometry of the outer region described by the line element (4.21).

At the critical time t_*, when $R_1(t_*) = R_2(t_*) = 1/(\sqrt{3}H(t_*))$, the normal to the surface of equation $\mathscr{F}(R) \equiv R - 1/(\sqrt{3}H(t_*)) = 0$ is $M_\mu \equiv \nabla_\mu \mathscr{F} = \delta_{1\mu}$ and

$$M^c M_c = g^{11} \left(R = \frac{1}{\sqrt{3}H(t_*)} \right) = \frac{2}{3} \left(1 - \sqrt{3}\, m_0 H(t_*) \right) = 0 \qquad (4.29)$$

and the instantaneous extremal apparent horizon is null.

One can compare the time rate of change of the apparent horizon radii with that of the cosmic substratum by differentiating Eq. (4.25) and solving for $\dot{R}_{AH}$, which yields

$$\dot{R}_{AH} = -\frac{2H\dot{H}R_{AH}^3}{3H^2 R_{AH}^2 - 1} \qquad (4.30)$$

and

$$\frac{\dot{R}_{AH}}{R_{AH}} - H = -H \left(1 + \frac{2\dot{H}R_{AH}^2}{3H^2 R_{AH}^2 - 1} \right). \qquad (4.31)$$

The apparent horizons are not comoving, except for trivial cases.[3] Even in a pure (spatially flat) FLRW universe (obtained in the limit $m = 0$), the cosmological apparent horizon at $R_{AH}(t) \equiv R_c(t) = 1/H(t)$ is not comoving, as discussed in Chap. 3. If an entropy can be ascribed to apparent horizons in General Relativity by the $S = \mathscr{A}/4$ prescription, then a natural question would be whether the total area of the McVittie apparent horizons is non-decreasing in time. The area $\mathscr{A}_1$ of the black hole apparent horizon is decreasing, but it is bounded from below by $16\pi m_0^2$, while this behaviour is more than compensated for by the increase of the area $\mathscr{A}_2$ of the cosmological apparent horizon. The total area

$$S = S_1 + S_2 = \pi \left(R_1^2 + R_2^2 \right) = \frac{\mathscr{A}}{4} = \frac{\mathscr{A}_1 + \mathscr{A}_2}{4} \qquad (4.32)$$

is plotted in Fig. 4.2.

Since the apparent horizons emerge as a pair at $t = t_*$, the horizon entropy S exhibits a discontinuous jump from zero value at this critical time.

---

[3] Also the known analytic solutions describing wormholes embedded in cosmological "backgrounds", which are few, show that these wormholes evolve in time [10, 50, 109, 160].

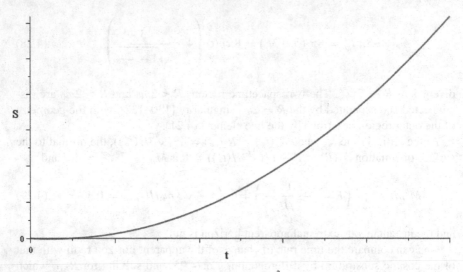

Fig. 4.2 The putative total horizon entropy S (in units $\dfrac{k_B c^3}{\hbar G}$, where k_B is the Boltzmann constant) associated with the apparent horizons as a function of time

4.3.2 Phantom McVittie Spacetime

Another possibility is that of a "background" FLRW universe sourced by a phantom fluid [55], defined by the equation of state parameter $w \equiv P/\rho < -1$ and violating the weak energy condition. Phantom fluids have raised much interest in relation with the present acceleration of the universe [5]. A notable feature is that they cause a Big Rip singularity at a finite time in the future $t_{\rm rip}$ [25]. The scale factor solving the Friedmann equation for a spatially flat FLRW universe with phantom scalar field is [25]

$$a(t) = \frac{a_0}{\left(t_{\rm rip} - t\right)^{\frac{2}{3|w+1|}}}, \qquad (4.33)$$

where $w < -1$, a_0, and $t_{\rm rip}$ are constants. The Hubble parameter

$$H(t) = \frac{2}{3|w+1|} \frac{1}{t_{\rm rip} - t}, \qquad (4.34)$$

is the time-reverse of that of a dust-dominated universe, $H(t) = 2/(3t)$. While the latter diverges at the Big Bang singularity and gradually decreases to zero, the former is finite at $t = 0$ and increases monotonically, diverging at the Big Rip. The McVittie apparent horizons describing black holes embedded in a FLRW phantom universe also behave as the time-reversal of those embedded in a FLRW "background" with $w > -1$, as shown in Fig. 4.3 [55].

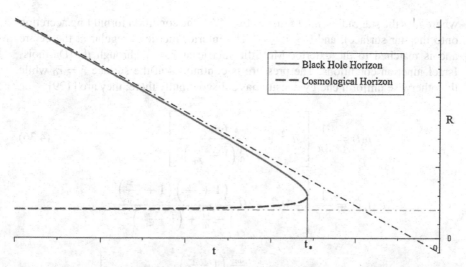

Fig. 4.3 The proper radii of the McVittie black hole (*continuous*) and cosmological (*dotted*) apparent horizons versus comoving time in a phantom-dominated universe. The parameter values are $w = -1.5$ and $t_{\rm rip} = 0$

The evolution of the apparent horizons in an expanding phantom universe proceeds as follows. Early on, there exist a black hole and a cosmological apparent horizon, which are approximately located at $R = 2m_0$ and $R = 1/H(t)$, respectively. As the universe grows older, the cosmological apparent horizon contracts while the black hole one expands, until the critical time t_* at which the two apparent horizons merge. At $t > t_*$ they both disappear exposing a naked singularity. The total apparent horizon area decreases and jumps discontinuously to zero value at the time t_*. This phenomenology is pretty bizarre, but it reflects the weirdness of the phantom fluid, which seems to violate the second law of thermodynamics in many ways [15, 70, 80, 104, 122, 127, 128] and may turn out to be completely unphysical.

4.3.3 Nolan Interior Solution

The Nolan interior solution [129] describes a relativistic star of uniform density in a FLRW "background" and provides, at least formally, a possible source for the McVittie metric. It mimics in the FLRW context the Schwarzschild interior solution with a Minkowski "background". The *Nolan line element* in isotropic coordinates is

$$ds^2 = -\left[\frac{1 - \frac{m}{r_0} + \frac{m\bar{r}^2}{\bar{r}_0^3}\left(1 - \frac{m}{4r_0}\right)}{\left(1 + \frac{m}{2r_0}\right)\left(1 + \frac{m\bar{r}^2}{2\bar{r}_0^3}\right)}\right]^2 dt^2 + a^2(t)\frac{\left(1 + \frac{m}{2\bar{r}}\right)^6}{\left(1 + \frac{m\bar{r}^2}{2\bar{r}_0^3}\right)^2}\left(d\bar{r}^2 + \bar{r}^2 d\Omega_{(2)}^2\right)$$

$$(4.35)$$

where $\bar{r}_0$ is the star radius, $m(t)$ satisfies Eq. (4.8) (the condition forbidding accretion onto the star surface), and $0 \leq \bar{r} \leq \bar{r}_0$. The interior metric is regular at the centre and is matched to the exterior McVittie metric at $\bar{r} = \bar{r}_0$ through the Darmois-Israel junction conditions. The pressure is continuous at the surface $\bar{r} = \bar{r}_0$ while the otherwise uniform energy density has a discontinuity there; they are [129]

$$\rho(t) = \frac{1}{8\pi} \left[3H^2 + \frac{6m}{a^2 \bar{r}_0^3 \left(1 + \frac{m}{2\bar{r}_0}\right)^6} \right], \qquad (4.36)$$

$$P(t, \bar{r}) = \frac{1}{8\pi} \left[-3H^2 - 2\dot{H} \frac{\left(1 + \frac{m}{2\bar{r}_0}\right)\left(1 + \frac{m\bar{r}^2}{2\bar{r}_0^3}\right)}{1 - \frac{m}{\bar{r}_0} + \left(1 - \frac{m}{4\bar{r}_0}\right)\frac{m\bar{r}^2}{\bar{r}_0^3}} \right.$$

$$\left. + \frac{\frac{3m^2}{\bar{r}_0^4}\left(1 - \frac{\bar{r}^2}{\bar{r}_0^2}\right)}{a^2 \left(1 + \frac{m}{2\bar{r}_0}\right)^6 \left[1 - \frac{m}{\bar{r}_0} + \left(1 - \frac{m}{4\bar{r}_0}\right)\frac{m\bar{r}^2}{\bar{r}_0^3}\right]} \right]. \qquad (4.37)$$

The Nolan interior solution (a member of the Kustaanheimo-Qvist family of shear-free solutions [93]) embeds the Schwarzschild interior solution with uniform constant density [169] in a time-dependent FLRW "background". The Schwarzschild interior solution is recovered if $a \equiv 1$. The energy density is positive-definite and the condition $P \geq 0$ imposed in Ref. [129] coincides with $\ddot{a} + 3\dot{a}^2/2 < 0$.

By continuity, if $\Sigma_0(t) = \{(t, \bar{r}, \theta, \varphi) : \bar{r} = \bar{r}_0\}$ is the star surface at time t, the metric on this 2-sphere must coincide with the restriction of the McVittie metric to this sphere

$$ds^2 \big|_{\Sigma_0} = -\frac{\left(1 - \frac{m(t)}{2\bar{r}_0}\right)^2}{\left(1 + \frac{m(t)}{2\bar{r}_0}\right)^2} dt^2 + a^2(t) \left(1 + \frac{m(t)}{2\bar{r}_0}\right)^4 \bar{r}_0^2 d\Omega_{(2)}^2 . \qquad (4.38)$$

The proper area of the star surface Σ_0 is

$$\mathscr{A}_{\Sigma_0}(t) = \int\int_{\Sigma_0} d\theta d\varphi \sqrt{g_{\Sigma_0}} = 4\pi a^2(t) \bar{r}_0^2 \left(1 + \frac{m(t)}{2\bar{r}_0}\right)^4 , \qquad (4.39)$$

where $g_{ab}\big|_{\Sigma_0}$ is the metric on Σ_0 at time t and g_{Σ_0} is its determinant. Upon use of the proper radius $R \equiv a(t)\bar{r}\left(1 + \frac{m}{2\bar{r}}\right)^2$, one finds $\mathscr{A}_{\Sigma_0}(t) = 4\pi R_0^2$. The star surface comoves with the cosmic fluid and has areal radius $R_0(t) = a(t)\bar{r}_0\left(1 + \frac{m}{2\bar{r}_0}\right)^2$.

The Nolan interior solution provides an example of a local, strongly gravitationally bound, relativistic object which is comoving: in this case the cosmic expansion prevails over the local dynamics.

Following [51], let us compute the generalization of the Tolman-Oppenheimer-Volkoff equation [169] for the Nolan model. The covariant conservation $\nabla^b T_{ab} = 0$ of a perfect fluid with energy-momentum tensor $T_{ab} = (P + \rho)\, u_a u_b + P g_{ab}$ splits into the two equations [169]

$$u^c \nabla_c \rho + (P + \rho)\, \nabla^c u_c = 0, \qquad (4.40)$$

$$h^c{}_a \partial_c P + (P + \rho)\, u^c \nabla_c u_a = 0, \qquad (4.41)$$

where u^a is the fluid 4-velocity and $h_{ab} \equiv g_{ab} + u_a u_b$ defines the projection operator $h_a{}^c$ onto the 3-space orthogonal to u^a. Since $u^\mu \propto \delta^{0\mu}$ in comoving coordinates and $u^c u_c = -1$, we have

$$u^\mu = u\, \delta^{0\mu} = \frac{\left(1 + \frac{m}{2\bar{r}_0}\right)\left(1 + \frac{m\bar{r}^2}{2\bar{r}_0^3}\right)}{1 - \frac{m}{\bar{r}_0} + \left(1 - \frac{m}{4\bar{r}_0}\right)\frac{m\bar{r}^2}{\bar{r}_0^3}} \cdot (1, 0, 0, 0) \qquad (4.42)$$

(or $u = |g_{00}|^{-1/2}$). Then Eqs. (4.40) and (4.8) give

$$\frac{\partial \rho}{\partial t} + 3H\,(P + \rho)\left\{1 - \frac{m}{\bar{r}_0}\left[\frac{3}{2\left(1 + \frac{m}{2\bar{r}_0}\right)} - \frac{\bar{r}^2}{2\bar{r}_0^2}\left(1 + \frac{m\bar{r}^2}{2\bar{r}_0^3}\right)^{-1}\right]\right\} = 0. \qquad (4.43)$$

Equation (4.43) is a more general form of the usual FLRW conservation equation $\dot{\rho} + 3H\,(P + \rho) = 0$, which is reobtained if m vanishes identically. There is no analogue of Eq. (4.43) for the Schwarzschild interior solution which has $H = 0$ and static energy density.

We can put Eq. (4.41) in the form

$$\partial_c P + u_c u^b \partial_b P + (P + \rho)\, u^b \nabla_b u_c = 0; \qquad (4.44)$$

for c set to 1, the computation of the covariant derivative yields

$$\frac{\partial P}{\partial r} + (P + \rho)\,\frac{m\bar{r}}{\bar{r}_0^3}\,\frac{\left(1 + \frac{m}{2\bar{r}_0}\right)}{\left(1 + \frac{m\bar{r}^2}{2\bar{r}_0^3}\right)\left[1 - \frac{m}{\bar{r}_0} + \frac{m\bar{r}^2}{\bar{r}_0^3}\left(1 - \frac{m}{4\bar{r}_0}\right)\right]} = 0. \qquad (4.45)$$

In the Newtonian limit $m/\bar{r}, m/\bar{r}_0 \ll 1$, $P \ll \rho$, $r \simeq \bar{r}$ [51], this equation reduces to

$$\frac{\partial P}{\partial r} + \frac{d\Phi_N}{dr}\,\rho = 0, \qquad (4.46)$$

where

$$\rho = m \left(\frac{4\pi}{3} r_0^3 \right)^{-1} \tag{4.47}$$

and

$$\Phi_N = \frac{m\bar{r}^2}{2\bar{r}_0^3} \tag{4.48}$$

is the Newtonian potential. The density (4.47) can also be obtained by setting the scale factor to unity in Eq. (4.36) and using the curvature radius. The equation of hydrostatic equilibrium receives the first order correction

$$\frac{dP}{dr} + \frac{d\Phi_N}{dr} \rho \left\{ 1 - \frac{3}{2} \left[\Phi_N(r) - \Phi_N(r_0) \right] \right\} = 0. \tag{4.49}$$

4.4 Charged McVittie Spacetime

A charged version of the original McVittie metric was found by Shah and Vaidya [150] and later generalized by Mashhoon and Partovi [115]. Charged and uncharged McVittie solutions are special cases of the Kustaanheimo-Qvist family [93]; related solutions and relevant research are reviewed in Chap. 4 of Ref. [92]. The charged McVittie solution was rediscovered by Gao and Zhang [64], who also generalized it to higher dimensions [65], and was studied in [56, 116, 117]. Conformal diagrams of the McVittie spacetime for various "backgrounds" were obtained in [38, 94, 95, 100].

Restricting again to a spatially flat FLRW "background", the charged McVittie line element and the only nonzero component of the Maxwell tensor are[4]

$$ds^2 = -\frac{\left[1 - \frac{(m_0^2 - Q^2)}{4a^2\bar{r}^2} \right]^2}{\left[\left(1 + \frac{m_0}{2a\bar{r}} \right)^2 - \frac{Q^2}{4a^2\bar{r}^2} \right]^2} dt^2$$

$$+ a^2(t) \left[\left(1 + \frac{m_0}{2a\bar{r}} \right)^2 - \frac{Q^2}{4a^2\bar{r}^2} \right]^2 \left(d\bar{r}^2 + \bar{r}^2 d\Omega_{(2)}^2 \right), \tag{4.50}$$

$$F^{01} = \frac{Q}{a^2\bar{r}^2 \left[1 - \frac{(m^2 - Q^2)}{4a^2\bar{r}^2} \right] \left[\left(1 + \frac{m_0}{2a\bar{r}} \right)^2 - \frac{Q^2}{4a^2\bar{r}^2} \right]^2}, \tag{4.51}$$

---

[4]A typographical error is present in the numerator of g_{00} in Ref. [64], but the line element appears correctly in the later references [65, 116, 117].

where $\bar{r}$ is the isotropic radius and $m_0 > 0$ and Q are the usual mass parameter and an electric charge parameter, respectively. If a is set to unity the line element (4.50) reduces to the Reissner-Nordström one in isotropic coordinates. As $\bar{r} \to +\infty$, it reduces again to the spatially flat FLRW line element, and if both m_0 and Q are set to zero, the spacetime is exactly the spatially flat FLRW one.

The areal radius is now

$$R(t, r) = a(t)\bar{r}\left[\left(1 + \frac{m_0}{2a(t)\bar{r}}\right)^2 - \frac{Q^2}{4a^2(t)\bar{r}^2}\right]$$

$$= m_0 + a(t)\bar{r} + \frac{m_0^2 - Q^2}{4a(t)\bar{r}}, \qquad (4.52)$$

with $R(t, \bar{r}) \geq m_0$ for $|Q| \leq m_0$. When $|Q| \leq m_0$, the function $R(t, \bar{r})$ decreases from $+\infty$ in the range $0 < a\bar{r} < \sqrt{m_0^2 - Q^2}/2$, reaches an absolute minimum

$$R_{\min} = m_0 + \sqrt{m_0^2 - Q^2}$$

at $a\bar{r} = \sqrt{m_0^2 - Q^2}/2$, and increases again to $+\infty$ for $a\bar{r} > \sqrt{m_0^2 - Q^2}/2$ because the isotropic radius corresponds to a double covering[5] of the spacetime region $R > m_0 + \sqrt{m_0^2 - Q^2} \geq m_0$. When $|Q| \geq m_0$, the areal radius R increases monotonically with $\bar{r}$ and the physical region $R \geq 0$ corresponds to $\bar{r} \geq \frac{|Q| - m_0}{2a(t)} \geq 0$.

In the following it will be useful to invert the relation between areal and isotropic radii, which yields

$$\bar{r}^2 - \frac{(R - m_0)}{a}\bar{r} + \left(\frac{m_0^2 - Q^2}{4a^2}\right) = 0. \qquad (4.53)$$

The positive root obeys

$$2a\bar{r} = R - m_0 + \sqrt{R^2 + Q^2 - 2m_0 R} \equiv f(R). \qquad (4.54)$$

The Ricci scalar is [56]

$$\mathcal{R} = \frac{6}{\left[1 - \frac{(m_0^2 - Q^2)}{4a^2\bar{r}^2}\right]}\left\{\frac{\ddot{a}}{a}\left[\left(1 + \frac{m_0}{2a\bar{r}}\right)^2 - \frac{Q^2}{4a^2\bar{r}^2}\right]\right.$$

$$\left. + H^2\left[1 - \frac{m_0}{a\bar{r}} + \frac{3\left(Q^2 - m_0^2\right)}{4a^2\bar{r}^2}\right]\right\}. \qquad (4.55)$$

---

[5]This fact is well known in the special case $a \equiv 1, Q = 0$ corresponding to the Schwarzschild spacetime [19, 170].

and the non-vanishing components of the Einstein tensor are[6]

$$G^t{}_t = -\frac{256\,Q^2 a^4 \bar{r}^4}{[m_0^2 - Q^2 + 4m_0 a\bar{r} + 4a^2\bar{r}^2]^4} + \frac{3\dot{a}^2}{a^2}\,, \tag{4.56}$$

$$G^{\bar{r}}{}_{\bar{r}} = \frac{256\,Q^2 a^4 \bar{r}^4}{[m_0^2 - Q^2 + 4m_0 a\bar{r} + 4a^2\bar{r}^2]^4}$$

$$+ \frac{\dot{a}^2(-5m_0^2 + 5Q^2 - 8m_0 a\bar{r} + 4a^2\bar{r}^2)}{a^2(-m_0^2 + Q^2 + 4a^2\bar{r}^2)}$$

$$+ \frac{2\ddot{a}(m_0^2 - Q^2 + 4m_0 a\bar{r} + 4a^2\bar{r}^2)}{a(-m_0^2 + Q^2 + 4a^2\bar{r}^2)}\,, \tag{4.57}$$

$$G^\theta{}_\theta = G^\varphi{}_\varphi = -\frac{256\,Q^2 a^4 \bar{r}^4}{[m_0^2 - Q^2 + 4m_0 a\bar{r} + 4a^2\bar{r}^2]^4}$$

$$+ \frac{\dot{a}^2(-5m_0^2 + 5Q^2 - 8m_0 a\bar{r} + 4a^2\bar{r}^2)}{a^2(-m_0^2 + Q^2 + 4a^2\bar{r}^2)}$$

$$+ \frac{2\ddot{a}(m_0^2 - Q^2 + 4m_0 a\bar{r} + 4a^2\bar{r}^2)}{a(-m_0^2 + Q^2 + 4a^2\bar{r}^2)}\,. \tag{4.58}$$

As in the McVittie case, one has $G_1^0 = 0$, which forbids the radial flow of cosmic fluid. For $m_0 = Q = 0$ Eq. (4.55) reduces to the Ricci scalar of the spatially flat FLRW universe $\mathscr{R} = 6\left(\dot{H} + 2H^2\right)$. For $a \equiv 1$ it reduces to zero (the Ricci scalar of the Reissner-Nordström metric: then the only matter source is the electromagnetic field with traceless energy-momentum tensor). If $|Q| \leq m_0$ there is a spacetime singularity at

$$a\bar{r} = \frac{\sqrt{m_0^2 - Q^2}}{2}\,, \tag{4.59}$$

corresponding to

$$R = m_0 + \sqrt{m_0^2 - Q^2}\,, \tag{4.60}$$

which divides again the spacetime into two disconnected parts. It occurs where the outer event horizon of the Reissner-Nordström spacetime would be without FLRW "background". In the extremal case $|Q| = m_0$ the singularity occurs at $\bar{r} = 0$ or $R = m_0$, again coinciding with the location of the outer event horizon of the Reissner-Nordström limit (the event horizon of the extremal Reissner-Nordström black hole).

---

[6]The Einstein tensor appearing in Refs. [117] and [116] misses a scale factor in the denominators.

This new spherical singularity is not present if $|Q| > m_0$, but then the invariant of the Maxwell tensor

$$F_{ab}F^{ab} = -\frac{Q^2}{a^2\bar{r}^4\left[\left(1+\frac{m_0}{2a\bar{r}}\right)^2 - \frac{Q^2}{4a^2\bar{r}^2}\right]^2} \tag{4.61}$$

is singular at

$$a\bar{r} = \frac{|Q| - m_0}{2}, \tag{4.62}$$

corresponding to $R = 0$, because the radial electric field (the only non-zero component F^{01} of F^{ab}) is singular. The Big Bang singularity $a = 0$ is also present.

The spacetime singularity (4.60) for $|Q| \le m_0$ is spacelike, because it is described by the equation $\psi = 0$, where

$$\psi\,(t,\bar{r}) \equiv a(t)\bar{r} - \frac{\sqrt{m_0^2 - Q^2}}{2}\,; \tag{4.63}$$

and

$$\nabla_c\psi\nabla^c\psi = -\dot{a}^2\bar{r}^2\frac{\left[\left(1+\frac{m_0}{2a\bar{r}}\right)^2 - \frac{Q^2}{4a^2\bar{r}^2}\right]^2}{\left[1 - \frac{(m_0^2 - Q^2)}{4a^2\bar{r}^2}\right]^2}$$
$$+ \frac{1}{\left[\left(1+\frac{m_0}{2a\bar{r}}\right)^2 - \frac{Q^2}{4a^2\bar{r}^2}\right]^2}\,; \tag{4.64}$$

in the limit $2a\bar{r} \to \left(\sqrt{m_0^2 - Q^2}\right)^+$, this expression tends to $-\infty$. The norm of the normal to the surfaces $\psi = $ constant is negative, and this surface and its limit are spacelike [56, 115].

The location of the apparent horizons is given by $\nabla^c R\nabla_c R = 0$, which reads

$$\left\{\dot{a}^2\bar{r}^2\left[\left(1+\frac{m_0}{2a\bar{r}}\right)^2 - \frac{Q^2}{4a^2\bar{r}^2}\right]^4 - \left[1 - \frac{(m_0^2 - Q^2)}{4a^2\bar{r}^2}\right]^2\right\}$$
$$\cdot\left[\left(1+\frac{m_0}{2a\bar{r}}\right)^2 - \frac{Q^2}{4a^2\bar{r}^2}\right]^{-2} = 0\,, \tag{4.65}$$

which gives

$$\dot{a}^2 \bar{r}^2 \left[\left(1 + \frac{m_0}{2a\bar{r}} \right)^2 - \frac{Q^2}{4a^2\bar{r}^2} \right]^4 = \left[1 - \frac{(m_0^2 - Q^2)}{4a^2\bar{r}^2} \right]^2 \tag{4.66}$$

or

$$4f^2 R^4 H^2 - \left(f^2 - m_0^2 + Q^2 \right)^2 = 0 , \tag{4.67}$$

with $f(R)$ as in Eq. (4.54). We then have to solve the quartic

$$H^2 R^4 - R^2 + 2m_0 R - Q^2 = 0 . \tag{4.68}$$

For large R, this equation reduces to the radius of the spatially flat FLRW "background" $R \approx H^{-1}$. In any regime in which $H \to 0$ (for example for a power-law scale factor $a(t) = a_0 t^p$), Eq. (4.67) becomes asymptotically

$$R^2 - 2m_0 R + Q^2 \simeq 0 \tag{4.69}$$

so that, in this limit

$$R = m_0 \pm \sqrt{m_0^2 - Q^2} . \tag{4.70}$$

In any region of a background for which $H \to 0$, a black hole apparent horizon asymptotes to the singularity (4.60). The smaller root is always covered by the spherical singularity and these two roots coincide with the locations of the two event horizons of the Reissner-Nordström spacetime. The limit $H \to 0$ reproduces the horizon structure of Reissner-Nordström (but with the spherical singularity added). This spherical singularity persists when $|Q| = m_0$ and is time-independent (as long as the scale factor a it is not exactly constant).

We restrict again to a dust-dominated FLRW substratum with $H(t) = \dfrac{2}{3t}$ and we begin with the case $|Q| < m_0$. Then Eq. (4.68) can be solved explicitly at different times t giving the location of the apparent horizons. The dynamics of the apparent horizon radii in comoving time is qualitatively the same that we have seen for the uncharged McVittie spacetime [55, 56, 85, 94, 95, 130, 131]. The innermost apparent horizon is located in the inner disconnected region, which is separated from the external geometry by the singularity (4.60) (which converged to $R = 0$ at the Big Bang). This innermost horizon asymptotes to the location of the inner unstable Cauchy horizon of the Reissner-Nordström geometry.

In the extremal case $|Q| = m_0$ the quartic (4.68) is solved exactly, with roots

$$R_{\text{extremal}} \equiv \frac{1}{2H} \pm \frac{\sqrt{1 - 4m_0 H}}{2H} , \quad \frac{-1}{2H} \pm \frac{\sqrt{1 + 4m_0 H}}{2H} . \tag{4.71}$$

Explicit analytic expressions for the apparent horizon radii are rare to find in studies of time-evolving black holes [49].

The root

$$R_* = -\frac{1}{2H} - \frac{\sqrt{1 + 4m_0 H}}{2H} \qquad (4.72)$$

is always negative and unphysical; the smallest positive root

$$R_{\text{inner}} \equiv -\frac{1}{2H} + \frac{\sqrt{1 + 4m_0 H}}{2H} \qquad (4.73)$$

always exists, while the other two roots

$$R_\pm \equiv \frac{1}{2H} \pm \frac{\sqrt{1 - 4m_0 H}}{2H} \qquad (4.74)$$

can merge (become complex) or appear simultaneously (become real) depending on the evolution of H. In a dust-dominated FLRW "background", H tends to zero at late times and R_{inner} and R_- in fact converge to the same radius $R = m_0$, which is the location of the single event horizon of the extremal Reissner-Nordström black hole.

In the supercritical case $|Q| > m_0$ we have a naked singularity, as expected since the limit $a \equiv 1$ reproduces the Reissner-Nordström spacetime. There is a cosmological horizon, given by the only root of Eq. (4.68) which is real and positive [56].

In the Reissner-Nordström black hole (to which the charged McVittie spacetime reduces if $a \equiv 1$) there are an outer event horizon and an inner (apparent) horizon. However, for the relevant range of parameters $|Q| \leq m_0$, there are only one black hole apparent horizon and one cosmological apparent horizon. This fact is interpreted as follows [56]. It is well known that the inner horizon of the Reissner-Nordström black hole is unstable with respect to linear perturbations [138] and the cosmological "background" perturbs the central inhomogeneity, only this is a non-linear (or exact) "large perturbation". It is conjectured in [56] that such an horizon will not appear in all exact solutions of General Relativity which describe a Reissner-Nordström black hole interacting with non-trivial environments.

4.5 An Application to the Quantization of Black Hole Areas

As an application, the charged McVittie spacetime was used[7] to disprove the universality of certain quantization laws for quantities constructed with the areas of black hole apparent horizons and inspired by string theories [53].

---

[7]The analysis of Ref. [53] makes use of the line element of [64] which contains an error but the qualitative behaviour of the apparent horizons for $|Q| \leq m_0$ does not change and the argument of Ref. [53], which is qualitative, is still valid.

The research community working on the holographic principle and stringy or supergravity black holes was excited by the observation that products of Killing horizon areas for certain multi-horizon black holes are independent of the black hole mass—they depend only on quantized charges (supergravity and extra-dimensional black holes with angular momentum and electric and magnetic charges were considered) [31, 35–37, 63, 99, 120]. This literature, inspired by the holographic principle and string theories (although the results are not strictly derived from string theories), betrays the idea that quantized products of areas depending on combinations of integers must carry the signature of some specific microphysics. If the entropy S of an horizon is given by the $S = \mathscr{A}/4$ prescription, statistical mechanics based on microscopic models counts microstates determined by quantum gravity and the horizon area should be quantized. When there are outer $(+)$ and inner $(-)$ horizons, the quantization rules appearing in the literature are

$$\mathscr{A}_{\pm} = 8\pi l_{\text{Pl}}^2 \left(\sqrt{N_1} - \sqrt{N_2} \right) \qquad N_1, N_2 \in \mathbb{N}, \qquad (4.75)$$

or

$$\mathscr{A}_+ \mathscr{A}_- = \left(8\pi l_{\text{Pl}}^2 \right)^2 N, \qquad N \in \mathbb{N}, \qquad (4.76)$$

where l_{Pl} is the Planck length [31, 35–37, 63, 99, 120]. $N_{1,2}$ and N are integers for supersymmetric extremal black holes, and depend on the number of branes, antibranes, and strings in more complicated situations [79]. A weaker rule states that the product of horizon areas is independent of the black hole mass and depends only on the quantized charges. These rules are often reported as universal ones valid for all black holes with multiple horizons. However, a warning about universality was issued by Visser [165, 166]. He studied black holes in 4-dimensional General Relativity and found that products of areas do not give mass-independent quantities, and they are not related in any simple way to integers. Instead, specific quadratic combinations of the various horizon radii (with the dimension of an area) generate mass-independent quantities and are, presumably, the best candidates to be quantized [165, 166]. It is essential to include in these algebraic combinations both cosmological and virtual horizons, in addition to physical black hole horizons [166]. (Virtual horizons are negative or imaginary roots of the equation locating the horizons.)

The horizons considered in the literature are Killing (and event) horizons. Realistic fundamental black holes cannot be stationary because, already at the semiclassical level, they emit Hawking radiation and the backreaction changes their masses, which become time-dependent, together with their horizon areas. For astrophysical black holes the effect is completely negligible but this cannot be the case for quantum black holes. Then, there will be no timelike Killing vector and apparent horizons should be considered instead of Killing and event horizons.

Visser's discussion of the Schwarzschild-de Sitter-Kottler black hole [165, 166] can be repeated almost without changes: since the calculations performed in these

works are algebraic, the only change is that the constant Hubble parameter of
the Schwarzschild-de Sitter-Kottler spacetime is replaced by the time-dependent
$H(t)$ of a FLRW "background". Including the virtual horizon in the count, it is
straightforward to see that the quantities

$$R_V (R_{BH} + R_C) + R_{BH} R_C = -\frac{1}{H^2(t)} \tag{4.77}$$

and

$$(R_{BH} + R_C)^2 - R_{BH} R_C = \frac{1}{H^2(t)} \tag{4.78}$$

are independent of the black hole mass m. This situation is a special case
[166]: mass-independent combinations of apparent horizon radii exist whenever
the Misner-Sharp-Hernandez mass is a Laurent polynomial of the areal radius R.
This is clearly the case of the uncharged and charged McVittie spacetimes. Taking
the McVittie spacetime as an example, the physical mass contained in a sphere is
the Misner-Sharp-Hernandez one and the cosmic fluid serves the only purpose of
generating a cosmological "background" to make the black hole dynamical. It seems
that the relevant mass to consider when mass-independent quantities such as (4.77)
and (4.78) are searched for is the black hole contribution m, not the total M_{MSH}. In
any case, the radii of the apparent horizons identify different spheres and correspond
to different Misner-Sharp-Hernandez masses $M_{MSH}^{(i)} = 2R_{AH}^{(i)}$. Here we stick to m.

Following [166], we include the virtual horizon to obtain mass-independent
quantities. Now, when the apparent horizons are time-dependent, the combinations (4.77) and (4.78) are also time-dependent. Even if they are expressed by
combinations of integers at an initial time, they will not be such immediately
afterward, and at all other times.

Although the cosmological black holes that we consider are just toy models for
non-stationary black hole horizons, the point is that realistic black holes are time-
dependent and far-reaching conclusions about quantizing black hole horizon areas,
or quantities quadratic in the radii of Killing horizons, are unwarranted. Generic
statements should be put on a firmer ground before being promoted to the role
of universal results. Our dynamical cosmological black hole examples reinforce
the argument of Refs. [165, 166] that the black holes of 4-dimensional General
Relativity do not reconcile with the quantization rules (4.75) and (4.76).

4.6 Generalized McVittie Spacetimes

The McVittie no-accretion condition can be relaxed to allow for generalized
McVittie solutions [51]. In the Synge approach, a metric can be prescribed and
forced to solve the Einstein equations by running the latter from left to right and by

computing the energy-momentum tensor corresponding to this solution. However, this stress-energy tensor usually turns out to be completely unphysical and to violate any reasonable energy condition, even the non-negativity of the energy density. It comes as a surprise that generalized McVittie solutions with relatively well-behaved matter sources exist.

Generalized McVittie solutions have line element

$$ds^2 = -\frac{B^2(t,\bar{r})}{A^2(t,\bar{r})}dt^2 + a^2(t)A^4(t,\bar{r})\left(d\bar{r}^2 + \bar{r}^2 d\Omega_{(2)}^2\right) \tag{4.79}$$

in isotropic coordinates, where $m(t)$ is a positive function of time and

$$A(t,\bar{r}) \equiv 1 + \frac{m(t)}{2\bar{r}}, \qquad B(t,\bar{r}) \equiv 1 - \frac{m(t)}{2\bar{r}}. \tag{4.80}$$

The Einstein tensor admits the only non-vanishing mixed components

$$G_0^0 = -\frac{3A^2}{B^2}\left(\frac{\dot{a}}{a} + \frac{\dot{m}}{\bar{r}A}\right)^2, \tag{4.81}$$

$$G_0^1 = \frac{2m}{a^2\bar{r}^2A^5B}\left(\frac{\dot{m}}{m} + \frac{\dot{a}}{a}\right), \tag{4.82}$$

$$G_1^1 = G_2^2 = G_3^3 = -\frac{A^2}{B^2}\left\{2\frac{d}{dt}\left(\frac{\dot{a}}{a} + \frac{\dot{m}}{\bar{r}A}\right) + \left(\frac{\dot{a}}{a} + \frac{\dot{m}}{\bar{r}A}\right)\right.$$
$$\left.\cdot\left[3\left(\frac{\dot{a}}{a} + \frac{\dot{m}}{\bar{r}A}\right) + \frac{2\dot{m}}{\bar{r}AB}\right]\right\}. \tag{4.83}$$

It is useful to consider the expression

$$C \equiv \frac{\dot{a}}{a} + \frac{\dot{m}}{\bar{r}A} = \frac{\dot{m}_H}{m_H} - \frac{\dot{m}}{m}\frac{B}{A} \tag{4.84}$$

which appears in the Einstein equations and reduces to $\dot{m}_H/m_H$, where

$$m_H \equiv m(t)a(t) \tag{4.85}$$

for the special subclass of solutions with $m = m_0 =$ constant referred to as "comoving mass" solutions (this class will be of some importance later). On the surface $\bar{r} = m/2$, C reduces to

$$C_\Sigma = \frac{\dot{a}}{a} + \frac{\dot{m}}{m} = \frac{\dot{m}_H}{m_H}. \tag{4.86}$$

McVittie solutions have $C_\Sigma = 0$, while comoving mass solutions are characterized by $C = C_\Sigma = H$ everywhere.

The Ricci scalar

$$\mathcal{R} = \frac{3A^2}{B^2}\left(2\dot{C} + 4C^2 + \frac{2\dot{m}C}{\bar{r}AB}\right) \tag{4.87}$$

diverges as $\bar{r} \to m/2$ except in the case when $m = $ constant. Distinct cosmic fluids or effective fluids can be contemplated as matter sources of generalized McVittie spacetimes, which we consider separately in what follows.

4.6.1 A Single Perfect Fluid

If the generalized McVittie metric is sourced by a single perfect fluid with stress energy tensor

$$T_{ab} = (P + \rho)\, u_a u_b + P g_{ab}, \tag{4.88}$$

a radial flow of cosmic fluid is described by the fluid 4-velocity $u^\mu = (u^0, u, 0, 0)$. Then, the only possible solution in General Relativity is the Schwarzschild-de Sitter-Kottler black hole [51, 66, 94]. In fact, using the normalization $u^c u_c = -1$, one finds

$$u^0 = \frac{A}{B}\sqrt{1 + a^2 A^4 u^2} \tag{4.89}$$

and, using Eqs. (4.81)–(4.83),

$$\dot{m}_H = -GB^2 au\,(P + \rho)\,\mathscr{A}\sqrt{1 + a^2 A^4 u^2}, \tag{4.90}$$

where

$$\mathscr{A} = \int\int d\theta d\varphi\,\sqrt{g_\Sigma} = 4\pi a^2 A^4 \bar{r}^2 \tag{4.91}$$

is the area of a sphere of isotropic radius $\bar{r}$ and

$$3\left(\frac{AC}{B}\right)^2 = 8\pi\left[(P + \rho)\,a^2 A^4 u^2 + \rho\right], \tag{4.92}$$

$$\left(\frac{A}{B}\right)^2\left(2\dot{C} + 3C^2 + \frac{2\dot{m}C}{\bar{r}AB}\right) = -8\pi\left[(P + \rho)\,a^2 A^4 u^2 + P\right], \tag{4.93}$$

$$\left(\frac{A}{B}\right)^2\left(2\dot{C} + 3C^2 + \frac{2\dot{m}C}{\bar{r}AB}\right) = -8\pi P. \tag{4.94}$$

Comparing Eqs. (4.93) and (4.94) gives $P = -\rho$ and the de Sitter equation of state is the unique possibility; Eq. (4.90) then implies $\dot{m}_H = 0$. A single perfect fluid cannot source the generalized McVittie spacetime (with the trivial exception of the non-accreting Schwarzschild-de Sitter-Kottler). A mixture of two perfect fluids constitutes a potential source.[8]

4.6.2   *Imperfect Fluid Without Radial Flow of Material*

Another possibility is to allow an imperfect fluid with energy-momentum tensor

$$T_{ab} = (P + \rho)\, u_a u_b + P g_{ab} + q_a u_b + q_b u_a \tag{4.95}$$

to source the generalized McVittie spacetime. The vector q^c is purely spatial[9] and describes a radial energy flow, but there is no flow of cosmic fluid:

$$u^\mu = \left(\frac{A}{B}, 0, 0, 0\right), \qquad q^\alpha = (0, q, 0, 0), \qquad q^c u_c = 0. \tag{4.96}$$

The $(0, 1)$ component of the Einstein equations is

$$\frac{\dot{m}}{m} + \frac{\dot{a}}{a} = -\frac{4\pi}{m} \bar{r}^2 a^2 A^4 B^2 q. \tag{4.97}$$

and

$$\frac{\dot{m}_H}{m_H} = \frac{\dot{m}}{m} + \frac{\dot{a}}{a}. \tag{4.98}$$

The area of a sphere Σ of constant time and isotropic radius is

$$\mathscr{A} = \int\!\!\int d\theta d\varphi\, \sqrt{g_\Sigma} = 4\pi a^2 A^4 \bar{r}^2,$$

and the relation between energy flow, area $\mathscr{A}$, and accretion rate

$$\dot{m}_H(t) = -aB^2 \mathscr{A} q \tag{4.99}$$

holds true. An energy inflow is described by $q < 0$ and this condition can be written on a sphere of radius $\bar{r} \gg m$ as

$$\dot{m}_H \simeq a \mathscr{A} |q|. \tag{4.100}$$

---

[8] A mixture of two perfect fluids is the matter source for the Sultana-Dyer solution (Sect. 4.7), which does not belong to the McVittie class.

[9] In principle, one could take this vector to be spacelike instead of purely spatial.

The $(0,0)$ and $(1,1)$ (or $(2,2)$ or $(3,3)$) components of the Einstein equations provide the energy density and pressure[10]

$$\rho(t,\bar{r}) = \frac{3A^2}{8\pi B^2}\left(\frac{\dot{a}}{a} + \frac{\dot{m}}{\bar{r}A}\right)^2,$$
(4.101)

$$P(t,\bar{r}) = \frac{-A^2}{8\pi B^2}\left\{2\frac{d}{dt}\left(\frac{\dot{a}}{a} + \frac{\dot{m}}{\bar{r}A}\right) + \left(\frac{\dot{a}}{a} + \frac{\dot{m}}{\bar{r}A}\right)\left[3\left(\frac{\dot{a}}{a} + \frac{\dot{m}}{\bar{r}A}\right) + \frac{2\dot{m}}{\bar{r}AB}\right]\right\}.$$
(4.102)

The energy density is always non-negative. The expansion scalar is $3C$ and Eq. (4.102) provides a generalized Raychaudhuri equation,

$$\dot{C} = -\frac{3C^2}{2} - \frac{\dot{m}}{\bar{r}AB}C - 4\pi\frac{B^2}{A^2}P,$$
(4.103)

which reduces to the usual equation of FLRW cosmology $\dot{H} = -\frac{3H^2}{2} - 4\pi P$ when $m \to 0$, and the Hamiltonian constraint $H^2 = 8\pi\rho/3$ then yields

$$\dot{H} = -4\pi(P + \rho).$$
(4.104)

In the general case, Eq. (4.101) gives the more general equation

$$\dot{C} = -4\pi\frac{B^2}{A^2}(P + \rho) - \frac{\dot{m}C}{\bar{r}AB}.$$
(4.105)

4.6.3 Imperfect Fluid with Radial Flow of Material

Yet another possibility is to have an imperfect fluid with energy-momentum tensor (4.95) and to allow for both radial fluid flow and an energy current:

$$u^\mu = \left(\frac{A}{B}\sqrt{1 + a^2A^4u^2}, u, 0, 0\right), \qquad q^\mu = (0, q, 0, 0).$$
(4.106)

The Einstein components (4.81)–(4.83) then give

$$\dot{m}_H = -aB^2\mathscr{A}\sqrt{1 + a^2A^4u^2}\,[(P + \rho)u + q],$$
(4.107)

$$\left(\frac{AC}{B}\right)^2 = \frac{8\pi}{3}\left[(P + \rho)a^2A^4u^2 + \rho\right],$$
(4.108)

---

[10]Contrary to the McVittie spacetime, now ρ depends also on the radial coordinate.

$$\left(\frac{A}{B}\right)^2 \left(2\dot{C} + 3C^2 + \frac{2\dot{m}C}{\bar{r}AB}\right) = -8\pi \left[(P+\rho)\,a^2A^4u^2 + P + 2a^2A^4qu\right],$$

$$(4.109)$$

$$\left(\frac{A}{B}\right)^2 \left(2\dot{C} + 3C^2 + \frac{2\dot{m}C}{\bar{r}AB}\right) = -8\pi P.$$

$$(4.110)$$

Subtracting the last two equations yields

$$q = -(P+\rho)\frac{u}{2},$$

$$(4.111)$$

that is, an ingoing radial flow of mass is associated with an outgoing radial heat current if $P > -\rho$. The substitution of Eq. (4.111) into (4.107) provides the accretion rate

$$\dot{m}_{\mathrm{H}} = -\frac{aB^2}{2}\sqrt{1 + a^2A^4u^2}\,(P+\rho)\,\mathscr{A}u,$$

$$(4.112)$$

where $(P+\rho)\,\mathscr{A}u$ is a flux of gravitating energy through the surface of area $\mathscr{A}$. Since $u < 0$ for inflow, m_{H} increases if $P + \rho > 0$, stays constant in a de Sitter "background", and decreases if phantom energy with $P < -\rho$ is accreted (which lends support to the findings of Ref. [6] about the fate of a black hole in a phantom universe[11]).

The energy density is

$$\rho = \frac{A^2}{8\pi B^2}\left[3C^2 + \frac{2a^2A^4u^2}{1 + a^2A^4u^2}\left(\dot{C} + \frac{\dot{m}C}{\bar{r}AB}\right)\right]$$

$$(4.113)$$

and, in the special case $m = m_0 = \text{constant}$, it reduces to

$$\rho = \frac{A^2}{8\pi B^2}\left[3H^2 + \frac{2\dot{H}a^2A^4u^2}{1 + a^2A^4u^2}\right].$$

$$(4.114)$$

It is positive in a superaccelerating universe with $\dot{H} > 0$. The velocity of the fluid is found to be

$$u = -\left\{\frac{\sqrt{1 + \frac{4m_0^2H^2a^2A^4}{B^4\mathscr{A}^2(P+\rho)^2}} - 1}{2a^2A^4}\right\}^{1/2}.$$

$$(4.115)$$

The fluid flow becomes superluminal as $\bar{r} \to m_0/2$, where $B \to 0$, an unphysical feature probably due to the oversimplified hydrodynamical model. In principle, it

---

[11]This study analyzes a test fluid in great detail and finds the same qualitative behaviour for the mass of a black hole accreting cosmic fluid.

does not make sense to study black holes in the presence of superluminal flows because, then, it is obvious that an horizon will not confine energy. However we restrict the flow to be inward, hence matter cannot flow superluminally outside of a black hole apparent horizon.[12] In reality, the matter/energy flow is subluminal and becomes supersonic at a certain radius, a feature which can only be modeled in a more realistic model of accretion.

4.6.4 "Comoving Mass" McVittie Solution

In the class of generalized McVittie solutions, the one corresponding to the choice $m_H(t) = m_0 a(t)$ is singled out because it is a late-time attractor *within this class* (Sect. 4.6.6). The *line element of the "comoving mass" McVittie solution* is

$$ds^2 = -\frac{\left(1 - \frac{m_0}{2\bar{r}}\right)^2}{\left(1 + \frac{m_0}{2\bar{r}}\right)^2} \, dt^2 + a^2(t) \left(1 + \frac{m_0}{2\bar{r}}\right)^4 \left(d\bar{r}^2 + r^2 d\Omega_{(2)}^2\right) \tag{4.116}$$

in isotropic coordinates. Using the radial coordinate $\tilde{r} \equiv \bar{r}\left(1 + \frac{m_0}{2\bar{r}}\right)^2$, the line element (4.116) becomes

$$ds^2 = -\left(1 - \frac{2m_0}{\tilde{r}}\right) dt^2 + a^2 \left(1 - \frac{2m_0}{\tilde{r}}\right)^{-1} d\tilde{r}^2 + a^2 \tilde{r}^2 d\Omega_{(2)}^2. \tag{4.117}$$

We now introduce the areal radius $R = a\tilde{r}$, obtaining

$$ds^2 = -\left[1 - \frac{2m_0 a}{R} - \left(1 - \frac{2m_0 a}{R}\right)^{-1} H^2 R^2\right] dt^2 + \left(1 - \frac{2m_0 a}{R}\right)^{-1} dR^2$$

$$-2HR\left(1 - \frac{2m_0 a}{R}\right)^{-1} dt\, dR + R^2 d\Omega_{(2)}^2 \tag{4.118}$$

and we eliminate the cross-term in $dtdR$ by introducing the new time T as [66]

$$dT = \frac{1}{F(t, R)}\left[dt + \frac{HR}{\left(1 - \frac{2m_0 a}{R}\right)^2 - H^2 R^2} dR\right], \tag{4.119}$$

where $F(t(T, R), R)$ is an integrating factor. The line element (4.118) assumes the form

---

[12]In principle energy can still flow superluminally inward across the cosmological horizon. The magnitude of the flux density q^c decreases with the radial distance from the black hole.

$$ds^2 = -\left[1 - \frac{2m_0 a}{R} - \left(1 - \frac{2m_0 a}{R}\right)^{-1} H^2 R^2\right] F^2 dT^2$$

$$+\left[1 - \frac{2m_0 a}{R} - \left(1 - \frac{2m_0 a}{R}\right)^{-1} H^2 R^2\right]^{-1} dR^2 + R^2 d\Omega^2_{(2)}, \quad (4.120)$$

where $H(t)$ and $F(t, R)$ are now functions of T and R. The areal radii of the apparent horizons, located by $g^{RR} = 0$, satisfy [66][13]

$$\mp HR^2 + R - 2m_0 a = 0. \quad (4.121)$$

Discarding the negative radius, there are two positive roots

$$R_{\rm C} = \frac{1}{2H}\left(1 + \sqrt{1 - 8m_0 \dot{a}}\right), \quad (4.122)$$

$$R_{\rm BH} = \frac{1}{2H}\left(1 - \sqrt{1 - 8m_0 \dot{a}}\right); \quad (4.123)$$

$R_{\rm C}$ is a cosmological and $R_{\rm BH}$ is a black hole apparent horizon. The singular surface $\tilde{r} = m_0/2$ corresponds to $\tilde{r} = 2m_0$ and to $R = 2m_0 a = 2m_{\rm H}(t)$ and $R_{\rm C,BH} > 2m_0 a = 2m_{\rm H}$.

The Misner-Sharp-Hernandez mass of the black hole is expressed analytically as

$$M_{\rm MSH}(t) = 2R_{\rm BH} = H^{-1}\left(1 - \sqrt{1 - 8m_0 \dot{a}}\right). \quad (4.124)$$

For a small black hole with $m_0 \dot{a} = \dot{m}_H \ll 1$, it is

$$M_{\rm MSH} \simeq 4m_0 \dot{a}(t) = 4\dot{m}_H(t). \quad (4.125)$$

4.6.5   More General Solutions

In the wider class of generalized McVittie solutions, the function $m(t)$ has arbitrary time dependence. The generalized McVittie line element is

$$ds^2 = -\left[1 - \frac{2m}{\tilde{r}} - \frac{a^2 \dot{m}^2}{1 - \frac{2m}{\tilde{r}}}\left(1 + \frac{m}{2\tilde{r}}\right)^2\right] dt^2 + a^2\left(1 - \frac{2m}{\tilde{r}}\right)^{-1} d\tilde{r}^2$$

$$+ a^2 \tilde{r}^2 d\Omega^2_{(2)} - \frac{2\dot{m}a^2\left(1 + \frac{m}{2r}\right)}{1 - \frac{2m}{\tilde{r}}} dt d\tilde{r}, \quad (4.126)$$

---

[13]This expression appears also in Ref. [43] and it can be derived also from Eq. (4.116) by expressing it in terms of R.

where $\tilde{r} = r\left(1 + \dfrac{M(t)}{2ra(t)}\right)^2$ and $m(t) \equiv M(t)/a(t)$. Rewriting Eq. (4.126) with the areal radius[14] $R \equiv a\tilde{r}$, one obtains

$$ds^2 = -\left[1 - \frac{2M}{R} - \frac{\left(HR + \dot{m}a\sqrt{\frac{\tilde{r}}{r}}\right)^2}{1 - \frac{2M}{R}}\right]dt^2 + \frac{dR^2}{1 - \frac{2M}{R}} + R^2 d\Omega_{(2)}^2$$

$$- \frac{2}{1 - \frac{2M}{R}}\left(HR + \dot{m}a\sqrt{\frac{\tilde{r}}{r}}\right)dt\,dR . \tag{4.127}$$

We introduce

$$A(t, R) \equiv 1 - \frac{2M}{R}, \tag{4.128}$$

$$\Gamma(t, R) \equiv HR + \dot{m}a\sqrt{\frac{\tilde{r}}{r}} \tag{4.129}$$

and we change from time t to the new time T which satisfies

$$dT = \frac{1}{F}\left(dt + \frac{C}{A^2 - \Gamma^2}\,dR\right), \tag{4.130}$$

with $F(T, R)$ an integrating factor, as usual. Then we have

$$ds^2 = -\left(1 - \frac{2M}{R}\right)\left[1 - \frac{\left(HR + \dot{m}a\sqrt{\frac{\tilde{r}}{r}}\right)^2}{\left(1 - \frac{2M}{R}\right)^2}\right]F^2 dT^2$$

$$+ \left(1 - \frac{2M}{R}\right)^{-1}\left[1 - \frac{\left(HR + \dot{m}a\sqrt{\frac{\tilde{r}}{r}}\right)^2}{\left(1 - \frac{2M}{R}\right)^2}\right]^{-1}dR^2$$

$$+ R^2 d\Omega_{(2)}^2 . \tag{4.131}$$

---

[14]$R(t, r)$ is an increasing function of r for $r > m/2$ since, in this range, $\frac{\partial R}{\partial r} = a\left(1 + \frac{M}{2ar}\right)\left(1 - \frac{M}{2ar}\right)$ is positive.

The usual recipe $g^{RR} = 0$ states that the apparent horizons are located at $\Gamma = \pm A$, which means that

$$HR + \dot{m}a\sqrt{\frac{\tilde{r}}{r}} = \pm\left(1 - \frac{2M}{R}\right). \tag{4.132}$$

It is

$$HR + M\left(1 + \frac{m}{2r}\right)\left(\frac{\dot{M}}{M} - H\right) = 1 - \frac{2M}{R}, \tag{4.133}$$

where the factor $M\left(1 + \frac{m}{2r}\right)$ quantifies the deviation of the radius from $2M$ (here $r > m/2$ is equivalent to $R > 2M$ and to $M\left(1 + \frac{2m}{r}\right) > 2M$). The quantity $\left(\frac{\dot{M}}{M} - H\right)$ is nothing but the difference between the percent rate of change of M and the corresponding rate of change of the "background" FLRW scale factor. When this quantity is zero, we have an analogue of the condition for stationary accretion, but now in a time-dependent "background" and under this condition the special solution with $M(t) = m_0 a(t)$ describes stationary accretion relative to the FRW "background".

Equation (4.132) becomes (excluding the negative root)

$$HR^2 + \left[M\left(1 + \frac{m}{2r}\right)\left(\frac{\dot{M}}{M} - H\right) - 1\right]R + 2M = 0. \tag{4.134}$$

Since now $M(t)$ does not scale as $a(t)$, we have the coefficient $\left(1 + \frac{m}{2r}\right)$, and Eq. (4.134) is not a second degree algebraic equation. However, one can decide to manipulate it (blindly) as a quadratic equation with formal "roots"

$$R_{\text{C,BH}} = \frac{1}{2H}\left\{1 - M\left(1 + \frac{m}{2r}\right)\frac{\dot{m}}{m} \pm \sqrt{\left[1 - M\left(1 + \frac{m}{2r}\right)\frac{\dot{m}}{m}\right]^2 - 8m\dot{a}}\right\}. \tag{4.135}$$

Because $r = r(R)$, Eq. (4.135) is an implicit equation for the radii $R_{\text{C,BH}}$ of the cosmological and black hole apparent horizons. If the argument of the square root in Eq. (4.135) is positive, there exist a cosmological apparent horizon (with proper radius R_C) and a black hole apparent horizon (with radius R_BH). If the argument of the square root is negative (which occurs near a Big Bang or a Big Rip singularity), there are no apparent horizons and there is a naked singularity embedded in a FLRW space. The critical situation corresponding to zero square root describes a moment of time at which the two apparent horizons coincide at $\sqrt{2M/H}$.

Conformal diagrams describing the global structure of the cosmological black holes depend on the FLRW "backgrounds": see [39, 66] for a discussion.

4.6.6 "Comoving Mass" Solution as an Attractor

The "comoving mass" solutions are generic under certain assumptions in the sense that at late times they are approached by all generalized McVittie solutions [54]. To prove this statement, one needs to assume that the universe always expands and that the function $m(t)$ is non-negative and is continuous with its first derivative.

Begin by writing Eq. (4.121) in terms of $\tilde{r} \equiv R/a$:

$$H\tilde{r} + \frac{2m}{\tilde{r}a} = -\dot{m}\left(1 + \frac{m}{2r}\right) + \frac{1}{a}. \tag{4.136}$$

Given that $m \geq 0$, the left hand side of this equation cannot be negative and

$$\dot{m}\left(1 + \frac{m}{2r}\right) < \frac{1}{a}. \tag{4.137}$$

Since $1 + \frac{m}{2r} > 0$ in an expanding universe in which $a \to +\infty$, it must be

$$\dot{m}_\infty \equiv \lim_{t \to +\infty} \dot{m}(t) \leq 0.$$

Now, if $\dot{m}_\infty = 0$, the function $m(t)$ becomes asymptotically comoving: i.e. $m(t) \approx m_0 a(t)$ for some positive constant m_0.

The other possibility is that $\dot{m}_\infty < 0$; then there is a time $\bar{t}$ such that, for all times $t > \bar{t}$, we have $\dot{m}(t) < 0$ and only two possibilities are left: since $m(t) \geq 0$, either $m(t)$ reaches the value zero at a finite time t_* with derivative $\dot{m}_* \equiv \dot{m}(t_*) < 0$, or $m(t) \to m_0 = $ constant with $\dot{m}(t) \to 0$ (that is, $m(t)$ has a horizontal asymptote).

If the first situation occurs, then at $t = t_*$ it is

$$HR = |\dot{m}_*|a + 1$$

and the radius of the black hole apparent horizon at t_* is

$$r_* \equiv r_{\text{horizon}}(t_*) = \frac{1}{H(t_*)}\left(|\dot{m}_*| + \frac{1}{a}\right). \tag{4.138}$$

Late in the history of the universe we have a black hole of zero mass $M(t_*) = a(t_*)m(t_*)$ but finite radius r_*. Evolving this situation in time generates a "black hole" with negative mass M and finite apparent horizon, a situation which is clearly unphysical and rules out the case $m(t_*) = 0$ with $m(t > t_*) < 0$ [54].

The remaining possibility is that $\dot{m}(t) \to 0$ at late times (which means $t \to +\infty$ in a universe which expands forever or $t \to t_{\text{rip}}$ if there is a Big Rip singularity at t_{rip}). The condition $\dot{m} \to 0$ means that at late times the rate of increase of $m(t)$ is at most the Hubble rate and this function becomes comoving (this conclusion does not hold at early times when $1/a$ in Eq. (4.136) does not tend to zero).

A trivial subcase of the second possibility, occurs if $m_0 \equiv 0$ and the solution reduces to a FLRW universe. Perhaps there is some merit in interpreting this situation as a black hole that evaporates completely, but the radial flow considered in these solutions is not described by a null vector.

Further, one can speculate that if the assumptions are relaxed, the black hole could avoid becoming comoving; for example, if $\dot{m}(t)$ is discontinuous, $m(t)$ could tend to zero in a finite time t_*, but this spacetime would have discontinuous connection coefficients and distributional curvature.

4.6.7   Recent Developments and Scalar Field Sources

The causal structure of the McVittie spacetime is rather complicated, and has been the subject of various recent studies [38, 85, 94, 130–132]. There is finally agreement that the McVittie metric describes what should be called a black hole when the black hole apparent horizon is present [94], but the motion of timelike and null test particles and the horizon structure depend heavily on the cosmological "background" [38].

It came as a surprise that scalar field sources for McVittie geometries are possible in various theories[15] [1, 3]. It was realized that the McVittie spacetime is not only a perfect fluid solution of the Einstein equations, but is also an analytic solution of a special form of k-essence called cuscuton [1] (k-essence theories have been originally formulated as dark energy models for cosmology, but since then they have taken a life of their own as possible fundamental theories). The McVittie spacetime admits constant mean curvature surfaces in its constant time foliation, and this fact makes the McVittie metric also a solution of Hořava-Lifschitz gravity (a theory very popular in the search for quantum gravity because of its renormalizability properties) with anisotropy. The McVittie solution is also an exact solution of shape dynamics [68] (another approach to quantum gravity). The cuscuton theory is the only form of k-essence which supports McVittie solutions [1].

In this optics, the generalized McVittie solution is also interesting, since it turns out to be an exact solution of Horndeski theory (the most general scalar-tensor theory admitting second order field equations) [3]. Rather than being a generalization of an old solution of General Relativity, a curiosity from the past, the McVittie and generalized McVittie solutions relate scalar fields and gravity in

---

[15]See Ref. [13] for scalar field sources of Lemaître-Tolman-Bondi models and the rest of this chapter for other scalar field solutions.

modern theories and allow research into the basic physics of black holes with scalar hair. This renaissance of the McVittie solution is still in its infancy and new results are expected in the near future.

4.7 Sultana-Dyer Spacetime

Another inhomogeneous and time-dependent solution of the Einstein equations which can be interpreted as a black hole embedded in a FLRW "background" universe is the Sultana-Dyer spacetime [158]. This solution is of Petrov type D and the "background" is a spatially flat FLRW universe evolving with the scale factor $a(t) = a_0 t^{2/3}$ of a dust fluid, however the matter source is a combination of a timelike dust and a null dust.

The Sultana-Dyer solution is the extension of the geometry generated by conformally transforming the Schwarzschild metric $g_{ab}^{(\text{Schw})}$:

$$g_{ab}^{(\text{Schw})} \rightarrow \Omega^2 g_{ab}^{(\text{Schw})}.$$ (4.139)

The conformal factor is

$$\Omega = a(\tau) = \tau^2,$$ (4.140)

where a is the scale factor of a $k = 0$ FLRW universe and τ is a time coordinate (which is neither the comoving nor the conformal time). The Sultana-Dyer metric is conformally static and possesses a conformal Killing vector ξ^a satisfying the conformal Killing equation [169]

$$\mathcal{L}_\xi g_{ab} = \nabla_a \xi_b + \nabla_b \xi_a = \frac{1}{2} g_{ab} \nabla^c \xi_c.$$ (4.141)

The original intention of Ref. [158] was to transform the timelike Killing field ξ^c of the Schwarzschild spacetime into a conformal Killing field (defined for $\xi^c \nabla_c \Omega \neq 0$), generating a conformal Killing horizon in the conformal cousin of the Schwarzschild spacetime. Nowadays, conformal Killing horizons seem to have little relevance in the study of evolving horizons, but the Sultana-Dyer spacetime remains a useful example.

The *Sultana-Dyer line element* is

$$ds^2 = a^2(\tau) \left[-\left(1 - \frac{2m_0}{r}\right) d\tau^2 + \frac{4m_0}{r} d\tau dr + \left(1 + \frac{2m_0}{r}\right) dr^2 + r^2 d\Omega_{(2)}^2 \right]$$ (4.142)

or

$$ds^2 = a^2(\tau) \left[-d\tau^2 + dr^2 + r^2 d\Omega_{(2)}^2 + \frac{2m_0}{r} (d\tau + dr)^2 \right],$$ (4.143)

where m_0 is a constant and $a(\tau) = \tau^2$. The coordinate change

$$\tau(t, r) = \eta + 2m_0 \ln \left| \frac{r}{2m_0} - 1 \right| \tag{4.144}$$

(where η is the conformal time of the FLRW "background") gives

$$d\tau = d\eta + \frac{2m_0}{r \left(1 - \frac{2m_0}{r}\right)} dr \tag{4.145}$$

which, substituted into Eq. (4.142), turns the line element into

$$ds^2 = a^2(\eta, r) \left[-\left(1 - \frac{2m_0}{r}\right) d\eta^2 + \frac{dr^2}{1 - \frac{2m_0}{r}} + r^2 d\Omega_{(2)}^2 \right]. \tag{4.146}$$

In this form, the metric is explicitly conformal to the Schwarzschild one with conformal factor

$$\Omega = a(\eta, r) = \tau^2(\eta, r) = \left(\eta + 2m_0 \ln \left| \frac{r}{2m_0} - 1 \right| \right)^2 \tag{4.147}$$

dependent on both η and r. The comoving time of the FLRW "background" is related to the conformal time η by $dt = a\,d\eta$. The Sultana-Dyer spacetime reduces to the spatially flat FLRW universe if $m \to 0$ or for $r \to +\infty$.

Isotropic coordinates are also used for the Sultana-Dyer metric. The isotropic radius $\bar{r}$ is defined by

$$r = \bar{r} \left(1 + \frac{m_0}{2\bar{r}}\right)^2. \tag{4.148}$$

Using this coordinate and the fact that

$$dr = \left(1 + \frac{m_0}{2\bar{r}}\right)\left(1 - \frac{m_0}{2\bar{r}}\right) d\bar{r}, \tag{4.149}$$

one obtains

$$ds^2 = a^2(\eta, r) \left[-\frac{\left(1 - \frac{m_0}{2\bar{r}}\right)^2}{\left(1 + \frac{m_0}{2\bar{r}}\right)^2} d\eta^2 + \left(1 + \frac{m_0}{2\bar{r}}\right)^4 \left(d\bar{r}^2 + \bar{r}^2 d\Omega_{(2)}^2 \right) \right]. \tag{4.150}$$

The matter source of the Sultana-Dyer spacetime is a combination of two non-interacting perfect fluids. The total energy-momentum tensor is

$$T_{ab} = T_{ab}^{(I)} + T_{ab}^{(II)} \equiv \rho u_a u_b + \rho_n k_a k_b, \tag{4.151}$$

where $T_{ab}^{(\mathrm{I})} = \rho u_a u_b$ is associated with an ordinary dust with timelike 4-velocity u^c, while $T_{ab}^{(\mathrm{II})} = \rho_n k_a k_b$ describes a null dust with density ρ_n and null vector k^c [158]. A problem pointed out already in the original work [158] is that in the Sultana-Dyer spacetime the cosmological fluid becomes tachyonic and the energy density becomes negative at late times near the spacetime singularity $\bar{r} = m_0/2$ [158].

The Misner-Sharp-Hernandez mass of a sphere of radius r in the Sultana-Dyer spacetime is easily computed

$$M_{\mathrm{MSH}} = m_0 a - 2m_0 r a_{,\tau} + \frac{r^3 a_{,\tau}^2}{2a}\left(1 + \frac{2m_0}{r}\right). \tag{4.152}$$

Using the identity

$$a_{,\eta} = a_{,\tau} = a\frac{\partial a}{\partial t} \equiv a\dot{a}, \tag{4.153}$$

the definition of the Hubble parameter $H \equiv \dot{a}/a$, it is easy to obtain [52]

$$M_{\mathrm{MSH}} = m_0\, a\, (1 - HR)^2 + \frac{H^2 R^3}{2}, \tag{4.154}$$

where $R = ar$ is the areal radius of the Sultana-Dyer spacetime. The mass M_{MSH} consists of two contributions: the first one is the mass m_0 of the Schwarzschild "seed" metric rescaled by the conformal factor a but scaled down by the cosmic expansion by $(1 - HR)^2$. This factor vanishes at $R = 1/H$, which is the radius of the cosmological horizon of the FLRW "background". The factors a and $(1 - HR)^2$ have competing effects which are not easy to interpret. The second contribution to M_{MSH} can be written as $\frac{4\pi}{3} R^3 \rho$, where $\rho = \frac{3H^2}{8\pi}$ is the density of the "background" cosmological fluid. This second term is obviously the mass of cosmic fluid contained in the sphere of areal radius R. Alternatively, one can write

$$M_{\mathrm{MSH}} = m_0 a + \frac{H^2 R^3}{2} - m_0 a HR(2 - HR) \equiv M_{\mathrm{MSH}}^{(1)} + M_{\mathrm{MSH}}^{(2)} + M_{\mathrm{MSH}}^{(\mathrm{int})}, \tag{4.155}$$

where the first term is purely local, the second one is purely cosmological, and the third one is an interaction term which is small for an object of size R much smaller than the Hubble radius H^{-1} of the "background".

The quantity

$$M(\bar{t}) \equiv m_0\, a(\bar{t}) \tag{4.156}$$

is not constant in the Sultana-Dyer solution. The locus $r = 2m_0$ is not a spacetime singularity; the conformal factor Ω diverges there but the original metric (4.142) is not singular and it can be considered as an extension of the conformally transformed Schwarzschild metric (4.146).

The Ricci curvature is

$$\mathscr{R} = \frac{12}{\eta^6} \left(1 - \frac{2m_0}{r} + \frac{2m_0\eta}{r^2} \right) , \tag{4.157}$$

and is not singular at $r = 2m_0$ (where $\eta \to -\infty$). There are spacetime singularities at $r = 0$ (central singularity) and at $\eta = 0$ (Big Bang).

To locate the apparent horizons[16] [52, 149] one must solve the equation $g^{RR} = 0$, where $R = ar$, obtaining

$$2m_0 a + \frac{r^3 a_{,\tau}^2}{a} \left(1 + \frac{2m_0}{r} \right) - 4m_0 r a_{,\tau} = ar . \tag{4.158}$$

Since $a(\tau) = \tau^2$, one obtains the cubic equation for r

$$4r^3 + 8m_0 r^2 - (8m_0 + \tau)\,\tau r + 2m_0\tau^2 = 0 , \tag{4.159}$$

the real positive roots of which are [52, 149]

$$r_1 = \frac{-4m_0 - \tau + \sqrt{\tau^2 + 24m_0\tau + 16m_0^2}}{4} , \tag{4.160}$$

$$r_2 = \frac{\tau(\eta, r)}{2} , \tag{4.161}$$

with $r_1 < r_2$. The surface $r = 2m_0$, which is the null event horizon of the Schwarzschild seed metric, remains an event horizon of the Sultana-Dyer metric. The radii of the apparent horizons expressed in terms of the areal radius $R = ar$ are

$$R_1 = \frac{\tau^3}{2} , \qquad R_2 = R_{\mathrm{EH}} = 2m_0\tau^2 , \tag{4.162}$$

and

$$R_4 = \frac{-4m_0 - \tau + \sqrt{\tau^2 + 24m_0\tau + 16m_0^2}}{4}\, \tau^2 . \tag{4.163}$$

These are implicit equations for the apparent horizons radii in terms of t and R.

The Sultana-Dyer spacetime was studied as an example of a time-dependent black hole horizon for which the Hawking temperature can be derived explicitly [149] to shed light on the Hawking effect and the thermodynamics in dynamical

---

[16]Beware of an error at the beginning of Ref. [48] consisting of imposing a coordinate condition which cannot be satisfied. This error was corrected in [29] and, later, in [159].

situations. In this context, the bad behaviour of matter near the horizon is not important. The authors of [149] studied the radiation of a massless, conformally coupled, scalar field ϕ in the Sultana-Dyer spacetime and computed the renormalized stress-energy tensor $\langle T_{ab} \rangle$ of ϕ. What makes the calculation feasible is the simplification due to the fact that the Sultana-Dyer geometry is conformal to the Schwarzschild one. The conformal anomaly and particle creation by the FLRW "background" were taken into account. Under the assumption that the Sultana-Dyer black hole evolves slowly, its effective Hawking temperature can be computed neglecting non-adiabatic terms. The result is [149]

$$\tilde{T} = \frac{1}{8\pi m_0 a} + \dots \tag{4.164}$$

where the ellipsis denotes corrections which are small in the limit of a slowly evolving black hole [149]. Since $T = \dfrac{1}{8\pi m_0}$ is the Hawking temperature of the "seed" Schwarzschild black hole generating the Sultana-Dyer geometry, one can write

$$\tilde{T} = \frac{T}{\Omega} + \dots \tag{4.165}$$

This result is a special case of a more general relation

$$T = \frac{T_{\text{Schw}}}{\Omega} \tag{4.166}$$

which is conjectured to hold [149] in spacetimes conformally related to the Schwarzschild spacetime by a transformation with conformal factor Ω. There is some independent support for this conjecture from naive dimensional considerations [47].

According to Dicke [42] (who followed earlier ideas of Weyl [171, 172]), a conformal transformation $g_{ab} \to \Omega^2 g_{ab}$ is nothing but a rescaling of the lengths of vectors and of the units used in a measurement, with the rescaling factor varying with the spacetime point. An experiment measures the ratio between a certain quantity x and its unit x_u. The quantity x itself is not meaningful unless a unit x_u is fixed for that quantity, and only the ratio x/x_u makes sense operationally. For example, the length of a ruler divided by the unit of length l_u is the same in the Minkowski metric η_{ab} and in a conformally related metric $g_{ab} = \Omega^2 \eta_{ab}$ if a new unit of length $\tilde{l}_u = \Omega l_u$ is associated with the length $\tilde{l}$ in a measurement. Two metrics g_{ab} and $\tilde{g}_{ab} = \Omega^2 g_{ab}$ are physically equivalent, at least from the classical point of view, when the units of the fundamental quantities length, time, and mass-energy are scaled according to $\tilde{l}_u = \Omega \, l_u$, $\tilde{t}_u = \Omega \, t_u$, and $\tilde{m}_u = \Omega^{-1} m_u$ [42] (units derived from the fundamental units are scaled according to their dimensions). There is no difference between using the Schwarzschild metric $g_{ab}^{(\text{Schw})}$ and the Sultana-Dyer metric conformal to it, $\tilde{g}_{ab} = g_{ab}^{(\text{SD})}$, provided that the units $\tilde{l}_u$, $\tilde{t}_u$, and $\tilde{m}_u$ are scaled appropriately (expanding for lengths and times, and redshifting for masses and energies). Since $k_B T$ is an energy, the ratio between $k_B T$ and m_u must be the same when using $g_{ab}^{(\text{Schw})}$ or $g_{ab}^{(\text{SD})}$:

$$\frac{k_B \tilde{T}}{\tilde{m}_u} = \frac{k_B T^{(\text{Schw})}}{m_u} . \tag{4.167}$$

Then the effective temperature of the Sultana-Dyer black hole follows,

$$\tilde{T} = \frac{T^{(\text{Schw})}}{\Omega} = \frac{1}{8\pi m_0 a} , \tag{4.168}$$

in agreement with Eq. (4.165).

In scalar-tensor gravity it is well-known that the transformation law of the stress-energy tensor of matter under conformal transformations $g_{ab} \to \tilde{g}_{ab} = \Omega^2 g_{ab}$ is

$$\tilde{T}_{ab}^{(\text{m})} = \Omega^{-2} T_{ab}^{(\text{m})} , \tag{4.169}$$

which agrees with a direct calculation of $\tilde{T}_{ab}^{(\text{m})}$ [46, 169]. By applying this rescaling to the semiclassical energy-momentum tensor of a scalar field near the Sultana-Dyer black hole, the renormalized $\langle \tilde{T}_{ab} \rangle$ should be

$$\langle \tilde{T}_{ab} \rangle = \frac{\langle T_{ab} \rangle}{a^2} . \tag{4.170}$$

The explicit renormalization of T_{ab} in Ref. [149] yields instead

$$\langle \tilde{T}_{ab} \rangle = \langle T_{ab}^{(\text{SD})} \rangle = \frac{\langle T_{ab} \rangle}{a^2} - \frac{1}{2880\pi^2} (X_{ab} - Y_{ab}) , \tag{4.171}$$

where

$$X_{ab} = 2\tilde{\nabla}_a \tilde{\nabla}_b \tilde{\mathscr{R}} - 2\tilde{g}_{ab} \tilde{\Box} \tilde{\mathscr{R}} + \frac{\tilde{\mathscr{R}}}{2} \tilde{g}_{ab} - 2\tilde{\mathscr{R}} \tilde{R}_{ab} , \tag{4.172}$$

$$Y_{ab} = -\tilde{R}_a^c \tilde{R}_{bc} + \frac{2}{3} \tilde{\mathscr{R}} \tilde{R}_{ab} + \frac{1}{2} \tilde{R}_{cd} \tilde{R}^{cd} \tilde{g}_{ab} - \frac{\tilde{\mathscr{R}}}{2} \tilde{g}_{ab} . \tag{4.173}$$

The extra terms in (4.171) are interpreted as quantum particle creation by the expanding FLRW "background" [149] (which cannot be predicted with Dicke's classical argument). When the black hole horizon evolves slowly, these terms can be neglected and the rescaling argument agrees with the $\langle \tilde{T}_{ab} \rangle$ of Eq. (4.170).

An independent argument in favour of Eq. (4.168) is the following [47]. The first law of black hole thermodynamics for a Schwarzschild black hole of mass m_0 is $TdS = dm_0$. The expression of the Bekenstein-Hawking entropy $S = \mathscr{A}/4$, where $\mathscr{A} = 4\pi r_{\text{H}}^2$ is the horizon area, together with the expression $r_{\text{H}} = 2m_0$ for the horizon radius, yields the Hawking temperature $T^{(\text{Schw})} = 1/\left(8\pi m_0\right)$. For a Sultana-Dyer expanding black hole, the purely local contribution to the Misner-Sharp-Hernandez mass is $M_{\text{MSH}}^{(1)} = m_0 a(t)$ and the proper horizon radius is $R_{\text{H}}(t) = a(t)r_{\text{H}}$. For this expanding horizon, the first law of black hole thermodynamics includes a work term dW:

$$TdS = dM_{\text{MSH}}^{(1)} + dW. \qquad (4.174)$$

If the black hole entropy is $S = \mathscr{A}/4$, then

$$8\pi T M_{\text{MSH}}^{(1)} dM_{\text{MSH}}^{(1)} = dM_{\text{MSH}}^{(1)} + dW. \qquad (4.175)$$

In the adiabatic approximation, the black hole should be in a state of quasi-equilibrium and the work terms should be negligible, which gives again

$$T \simeq \frac{1}{8\pi M_{\text{MSH}}(t)} = \frac{T^{(\text{Schw})}}{a}. \qquad (4.176)$$

The conformal factor of the Sultana-Dyer black hole does not depend on the radial coordinate and, in the adiabatic approximation in which its time variation is small, the Hawking temperature does have the scaling behaviour expected on dimensional grounds. This scaling law, however, will break down as soon as the conformal transformation is allowed to be radial-dependent, or the apparent horizon is allowed to vary rapidly.

The temperature of the Sultana-Dyer apparent horizon was calculated in [110] using the method of chiral anomalies. The result confirms the calculation of [149] and the guess of [47]. Moreover, the temperature is related to the entropy and the Misner-Sharp-Hernandez mass by the algebraic expression $M_{\text{MSH}} = 2ST$, which is the Smarr formula for stationary black holes.

4.8   Husain-Martinez-Nuñez Spacetime

The Husain-Martinez-Nuñez spacetime [82] is a solution of the Einstein equations which shows us a dynamics of the apparent horizons which is different from what we have encountered thus far. This spacetime is inhomogeneous, with a spatially flat FLRW "background" and the matter source is not a fluid but a free scalar field minimally coupled with gravity. The coupled Einstein-Klein-Gordon equations are

$$R_{ab} - \frac{1}{2} g_{ab} R = 8\pi T_{ab}^{(\phi)}, \qquad (4.177)$$

$$\Box \phi = 0, \qquad (4.178)$$

where

$$T_{ab}^{(\phi)} = \nabla_a\phi\nabla_b\phi - \frac{1}{2}g_{ab}\nabla_c\phi\nabla^c\phi.$$ (4.179)

Equation (4.177) simplifies to

$$R_{ab} = 8\pi\nabla_a\phi\nabla_b\phi.$$ (4.180)

The line element and scalar field are presented as [82]

$$ds^2 = (A_0\eta + B_0)\left[-\left(1 - \frac{2C}{r}\right)^\alpha d\eta^2 + \frac{dr^2}{\left(1 - \frac{2C}{r}\right)^\alpha}\right.$$

$$\left. +r^2\left(1 - \frac{2C}{r}\right)^{1-\alpha}d\Omega_{(2)}^2\right],$$ (4.181)

$$\phi(\eta, r) = \pm\frac{1}{4\sqrt{\pi}}\ln\left[D\left(1 - \frac{2C}{r}\right)^{\alpha/\sqrt{3}}(A_0\eta + B_0)^{\sqrt{3}}\right],$$ (4.182)

where $A_0, B_0, C,$ and D are non-negative constants, $\alpha = \pm\sqrt{3}/2$ is a parameter which can assume only two possible values, and $\eta > 0$. The additive constant B_0 can be dropped if $A_0 \neq 0$ because, in this case, it is redundant. If $A_0 = 0$ the Husain-Martinez-Nuñez line element (4.181) reduces to that of the Fisher spacetime [61]

$$ds^2 = -V^\nu(r)\,d\eta^2 + \frac{dr^2}{V^\nu(r)} + r^2V^{1-\nu}(r)d\Omega_{(2)}^2,$$ (4.183)

where $V(r) = 1 - 2\mu/r$, ν is a dimensionless parameter, μ is a mass, and the scalar field is

$$\psi(r) = \psi_0\ln V(r).$$ (4.184)

The static Fisher solution of the Einstein-Klein-Gordon equations is better known as the Janis-Newman-Winicour or the Wyman solution because it has been redis-covered again and again over the years and has picked up several names [4, 9, 18, 84, 164, 175]. The Fisher spacetime is asymptotically flat and contains a naked singularity at $r = 2C$. It is identified with the most general static spherically symmetric solution of the Einstein equations with zero cosmological constant and a massless minimally coupled scalar field as the source [146], but it is unstable [2]. The Husain-Martinez-Nuñez line element is conformal to the Fisher one, with the scale factor of the "background" FLRW space as the conformal factor, $\Omega = \sqrt{A_0\eta + B_0}$, and with only two possible values of the Fisher parameter ν. In the following we set $B_0 = 0$ and we denote the time of the Big Bang singularity with $\eta = 0$. The sign appearing in Eq. (4.182) is independent of the sign of α. As

$r \to +\infty$ the metric (4.181) is asymptotically FLRW, and it is FLRW if $C = 0$ (then the constant A_0 can be eliminated by rescaling η).

It is straightforward to compute the Ricci scalar as

$$\mathscr{R} = 8\pi \nabla^c \phi \nabla_c \phi = \frac{2\alpha^2 C^2 \left(1 - \frac{2C}{r}\right)^{\alpha-2}}{3r^4 A_0 \eta} - \frac{3A_0^2}{2 \left(A_0 \eta\right)^3 \left(1 - \frac{2C}{r}\right)^{\alpha}} . \tag{4.185}$$

Equation (4.185) shows that there is a spacetime singularity at $r = 2C$ for both values of α. The scalar field ϕ also diverges there (in addition, a Big Bang singularity occurs at $\eta = 0$). Only the range $2C < r < +\infty$ is physically meaningful. The areal radius is

$$R(\eta, r) = \sqrt{A_0 \eta}\, r \left(1 - \frac{2C}{r}\right)^{\frac{1-\alpha}{2}} \tag{4.186}$$

and the value $r = 2C$ of the radial coordinate corresponds to $R = 0$,

It is physically more rewarding to express the Husain-Martinez-Nuñez line element in terms of the comoving time t of the "background" FLRW space defined by $dt = a d\eta$, where $a(\eta) = \sqrt{A_0 \eta}$ is the FLRW scale factor. Since

$$t = \int d\eta\, a(\eta) = \frac{2\sqrt{A_0}}{3} \eta^{3/2} \tag{4.187}$$

fixing $\eta = 0$ at $t = 0$, it is

$$\eta = \left(\frac{3}{2\sqrt{A_0}} t\right)^{2/3} \tag{4.188}$$

and

$$a(t) = \sqrt{A_0 \eta} = a_0 t^{1/3}, \qquad a_0 = \left(\frac{3A_0}{2}\right)^{1/3}. \tag{4.189}$$

This scale factor is, of course, the one dictated by the stiff equation of state $P = \rho$ of a free massless scalar field in a FLRW universe. The general FLRW solution[17] for equation of state parameter $w \equiv P/\rho$ is

$$a(t) = \text{const.}\, t^{\frac{2}{3(w+1)}} . \tag{4.190}$$

---

[17]In a FLRW universe there are no spatial scalar field gradients (which would identify a preferred spatial direction) and the energy density and pressure are simply $\rho^{(\phi)} = \frac{\dot{\phi}^2}{2} + V(\phi)$, $P^{(\phi)} = \frac{\dot{\phi}^2}{2} - V(\phi)$. If $V(\phi) = 0$, then it is $P^{(\phi)} = \rho^{(\phi)}$.

The *Husain-Martinez-Nuñez line element* in comoving time is

$$ds^2 = -\left(1 - \frac{2C}{r}\right)^{\alpha} dt^2 + a^2(t)\left[\frac{dr^2}{\left(1 - \frac{2C}{r}\right)^{\alpha}} + r^2\left(1 - \frac{2C}{r}\right)^{1-\alpha} d\Omega_{(2)}^2\right] \quad (4.191)$$

and the scalar field sourcing it is

$$\phi(t,r) = \pm\frac{1}{4\sqrt{\pi}} \ln\left[D\left(1 - \frac{2C}{r}\right)^{\alpha/\sqrt{3}} a^{2\sqrt{3}}(t)\right]. \quad (4.192)$$

The areal radius (4.186) increases with r for $r > 2C$. We now recast the Husain-Martinez-Nuñez metric using the areal radius R. Using the notation

$$A(r) \equiv 1 - \frac{2C}{r}, \qquad B(r) \equiv 1 - \frac{(\alpha + 1)C}{r}, \quad (4.193)$$

one has $R(r) = a(t)rA^{\frac{1-\alpha}{2}}(r)$ and

$$dr = \left[A^{\frac{\alpha+1}{2}}\frac{dR}{a} - AH\,r\,dt\right]\frac{1}{B(r)}, \quad (4.194)$$

which leads to

$$ds^2 = -A^{\alpha}\left[1 - \frac{H^2R^2A^{2(1-\alpha)}}{B^2(r)}\right]dt^2 + \frac{H^2R^2A^{2-\alpha}(r)}{B^2(r)}dR^2$$

$$- \frac{2HRA^{\frac{3-\alpha}{2}}}{B^2(r)}dt\,dR + R^2d\Omega_{(2)}^2. \quad (4.195)$$

The time-radius cross-term is eliminated by using a new time T given by

$$dT = \frac{1}{F}(dt + \beta dR), \quad (4.196)$$

where $\beta(t,R)$ is a function to be determined and $F(t,R)$ is an integrating factor obeying the usual equation

$$\frac{\partial}{\partial R}\left(\frac{1}{F}\right) = \frac{\partial}{\partial t}\left(\frac{\beta}{F}\right). \quad (4.197)$$

Using $dt = FdT - \beta dR$ in Eq. (4.195) and choosing

$$\beta(t,R) = \frac{HRA^{\frac{3(1-\alpha)}{2}}}{B^2(r) - H^2R^2A^{2(1-\alpha)}}, \quad (4.198)$$

the line element becomes

$$ds^2 = -A^\alpha(r) \left[1 - \frac{H^2 R^2 A^{2(1-\alpha)}(r)}{B^2(r)} \right] F^2 dt^2$$
$$+ \frac{H^2 R^2 A^{2-\alpha}(r)}{B^2(r)} \left[1 + \frac{A^{1-\alpha}(r)}{B^2(r) - H^2 R^2 A^{2(1-\alpha)}(r)} \right] dR^2 + R^2 d\Omega_{(2)}^2 .$$

$$(4.199)$$

The apparent horizons, located by $g^{RR} = 0$, solve

$$B(r) = H(t) R A^{1-\alpha}(r) , \qquad (4.200)$$

where $r = r(t, R)$. In terms of the original coordinates η and r one has [82]

$$\frac{1}{\eta} = \frac{2}{r^2} \left[r - (\alpha + 1)C \right] \left(1 - \frac{2C}{r} \right)^{\alpha-1} . \qquad (4.201)$$

For $r \to +\infty$ (or $R \to +\infty$), Eq. (4.201) reduces to $R \simeq H^{-1}$, which is the radius of the cosmological apparent horizon of spatially flat FLRW space. Equation (4.200) can only be solved numerically. Let $x \equiv C/r$, then the equation locating the apparent horizons is

$$HR = \left[1 - \frac{(\alpha + 1)C}{r} \right] \left(1 - \frac{2C}{r} \right)^{\alpha-1} . \qquad (4.202)$$

The left hand side can be written as

$$HR = \frac{a_0}{3\, t^{2/3}} \frac{2C}{x} (1 - 2x)^{\frac{1-\alpha}{2}}$$

expressing the radius of the apparent horizons in units of H^{-1}; this is the radius of the cosmological apparent horizon of the FLRW "background". The right hand side of Eq. (4.202) is $[1 - (\alpha + 1)x] (1 - 2x)^{\alpha-1}$, hence Eq. (4.202) and the equation defining the areal radius give

$$t(x) = \left\{ \frac{2Ca_0}{3} \frac{(1 - 2x)^{3(1-\alpha)}}{x \left[1 - (\alpha + 1)x \right]} \right\}^{3/2} , \qquad (4.203)$$

$$R(x) = a_0\, t^{1/3}(x) \frac{2C}{x} (1 - 2x)^{\frac{1-\alpha}{2}} , \qquad (4.204)$$

respectively, which is a parametric representation of the function $R(t)$ and is then used to plot this function in Figs. 4.4 and 4.5. If the parameter value $\alpha = \sqrt{3}/2$ is adopted, in the time interval between the Big Bang and a critical time t_* there is

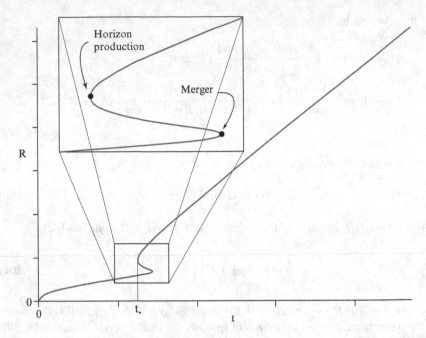

Fig. 4.4 The radii of the Husain-Martinez-Nuñez apparent horizons versus comoving time for $\alpha = \sqrt{3}/2$. Time t and radius R are both measured in arbitrary units of length and the parameters C and a_0 are chosen so that $(Ca_0)^{3/2} = 10^3$ in Eq. (4.203)

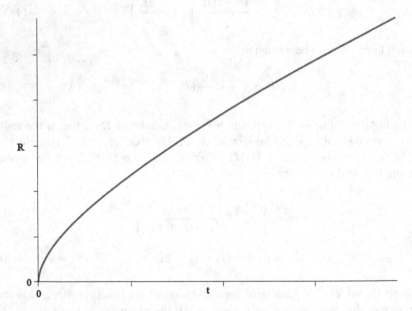

Fig. 4.5 The areal radius of the single Husain-Martinez-Nuñez apparent horizon present for the parameter value $\alpha = -\sqrt{3}/2$ versus comoving time. This cosmological apparent horizon expands and there is always a naked singularity at $R = 0$

only one apparent horizon which expands. At the time t_*, two additional apparent horizons appear; one is a cosmological apparent horizon which expands forever, while the other is a black hole horizon which contracts until it merges with the first apparent horizon which has been growing in the meantime. When they encounter each other, these two apparent horizons merge and disappear, causing a naked singularity at $R = 0$ to be present in a FLRW universe for the rest of the time. This apparent horizons dynamics differs from that of the McVittie and generalized McVittie solutions. The "S-curve" of Fig. 4.4 is recurrent in analytic solutions of Brans-Dicke and $f(\mathscr{R})$ gravity (Chap. 5). The scalar field is regular on the apparent horizons.

For the parameter value $\alpha = -\sqrt{3}/2$ only one cosmological apparent horizon exists at all times, and it expands. A naked singularity is present at $R = 0$ (Fig. 4.5), in addition to the usual spacelike Big Bang singularity at $t = 0$.

For the Husain-Martinez-Nuñez spacetime, it is possible to establish that the apparent horizons are *spacelike* by studing the normal vector to these surfaces and seeing that this vector always lies inside the light cone in an (η, r) diagram (we follow Ref. [82] here). According to Eq. (4.201), along the apparent horizons it is

$$\eta = \frac{r^2 \left(1 - \frac{2C}{r}\right)^{1-\alpha}}{2\left[r - C(1 + \alpha)\right]}. \tag{4.205}$$

Differentiating with respect to r, we obtain

$$\eta_{,r}\Big|_{\text{AH}} = \left(1 - \frac{2C}{r}\right)^{-\alpha} \left\{1 - \frac{r^2 \left(1 - \frac{2C}{r}\right)}{2\left[r - C(1 + \alpha)\right]^2}\right\}. \tag{4.206}$$

By comparison, along radial null geodesics, we have

$$\eta_{,r}\Big|_{\text{light cone}} = \pm \left(1 - \frac{2C}{r}\right)^{-\alpha} \tag{4.207}$$

(which can be obtained by setting $ds^2 = 0$ together with $d\theta = d\varphi = 0$), so that [82]

$$\left|\frac{\eta_{,r}\big|_{\text{AH}}}{\eta_{,r}\big|_{\text{light cone}}}\right| = 1 - \frac{\left(1 - \frac{2C}{r}\right)}{2\left[1 - \frac{(\alpha+1)C}{r}\right]^2} \leq 1. \tag{4.208}$$

The normal to the apparent horizon is always enclosed by the light cone, hence it is timelike, except where this vector becomes tangent to the light cone itself and it is null (which occurs when a pair of apparent horizons is created or disappears) [82].

The nature of the singularity at $r = 2C$ (or $R = 0$) is easily assessed. All surfaces described by $f(R) \equiv R - \text{const.} = 0$ have gradient $N_\mu \equiv \nabla_\mu f = \delta_{\mu 1}$ in (t, R, θ, φ)

coordinates. The norm squared of this gradient is

$$N_c N^c = g^{11} = \frac{B^2}{H^2 R^2 A^{2-\alpha}} \frac{1}{1 + \frac{A^{1-\alpha}}{B^2 - H^2 R^2 A^{2(1-\alpha)}}}. \tag{4.209}$$

Now, $B(r) \to \frac{1-\alpha}{2}$ and $A(r) \to 0^+$ as $r \to 2C^+$, therefore $N_c N^c > 0$ and $N_c N^c \to +\infty$ as $r \to 2C^+$. The $R = 0$ singularity is timelike for both values of the parameter α.

The Husain-Martinez-Nuñez spacetime is quoted as describing scalar field collapse, but a better description (for the parameter value $\alpha = +\sqrt{3}/2$) is that it exhibits the creation and annihilation of pairs of black hole apparent horizons. The $R = 0$ singularity (for both values of α) is created with the universe in the Big Bang and is not the product of gravitational collapse.

According to what already seen in this chapter, the physical interpretation of the apparent horizon dynamics for $\alpha = +\sqrt{3}$ is that a black hole larger than the cosmological horizon cannot fit in the early "universe". When this "universe" becomes sufficiently large, a black hole appears with inner and outer apparent horizons. These black hole horizons then merge into an extremal (null) black hole horizon and disappear.

4.9 Fonarev Solutions

The *Fonarev spacetime* of General Relativity has as the matter source a minimally coupled scalar field with an exponential self-interaction potential [62]. It describes a central inhomogeneity in an otherwise FLRW universe. The theory is described by the action

$$S = \int d^4 x \sqrt{-g} \left[\mathscr{R} - \frac{1}{2} \nabla_a \phi \nabla^a \phi - V(\phi) \right], \tag{4.210}$$

where

$$V(\phi) = V_0 \, e^{-\lambda \phi}, \tag{4.211}$$

with λ and V_0 positive constants. This potential has been investigated at length in FLRW cosmology [34, 73, 86, 103, 111]. The coupled Einstein-Klein-Gordon equations are

$$G_{ab} = 8\pi \left[\nabla_a \phi \nabla_b \phi - \frac{1}{2} g_{ab} \nabla_c \phi \nabla^c \phi - g_{ab} V(\phi) \right], \tag{4.212}$$

$$\Box\phi - \frac{dV}{d\phi} = 0 \,. \tag{4.213}$$

Equation (4.212) can be simplified to

$$R_{ab} = 8\pi \left(\nabla_a \phi \nabla_b \phi + g_{ab} V \right) \,. \tag{4.214}$$

The spherically symmetric Fonarev line element and scalar field are

$$ds^2 = a^2(\eta) \left[-f^2(r)\, d\eta^2 + \frac{dr^2}{f^2(r)} + S^2(r)\, d\Omega_{(2)}^2 \right] , \tag{4.215}$$

$$\phi(\eta, r) = \frac{1}{\sqrt{\lambda^2 + 2}} \ln\left(1 - \frac{2w}{r} \right) + \lambda \ln a + \frac{1}{\lambda} \ln\left[\frac{V_0 \left(\lambda^2 - 2 \right)^2}{2 A_0^2 \left(6 - \lambda^2 \right)} \right] , \tag{4.216}$$

where

$$f(r) = \left(1 - \frac{2w}{r} \right)^{\frac{\alpha}{2}} , \quad \alpha = \frac{\lambda}{\sqrt{\lambda^2 + 2}} , \tag{4.217}$$

$$S(r) = r \left(1 - \frac{2w}{r} \right)^{\frac{1-\alpha}{2}} , \quad a(\eta) = A_0 |\eta|^{\frac{2}{\lambda^2 - 2}} , \tag{4.218}$$

with w and A_0 constants. For simplicity we choose $A_0 = 1$. η is the conformal time of the FLRW "background". When $w = 0$ the metric (4.215) reduces to a spatially flat FLRW one while, when $a \equiv 1$ and $\alpha = 1$, it degenerates into the Schwarzschild solution (however, the value $\alpha = 1$ cannot be obtained if the condition $\alpha = \dfrac{\lambda}{\sqrt{\lambda^2 + 2}}$ holds). The line element becomes asymptotically that of a spatially flat FLRW space as $r \to +\infty$ (see the end of this section for a discussion of the apparent horizons).

The Fonarev metric is clearly a generalization of the Husain-Martinez-Nuñez class of solutions (4.181) to exponential potentials. The Husain-Martinez-Nuñez metric is recovered by setting $\lambda = \pm\sqrt{6}$ and $V_0 = 0$.

A *phantom Fonarev solution* corresponding to a dynamical phantom scalar field solution of General Relativity was introduced in Ref. [66]. It is obtained from the Fonarev solution using the transformation

$$\phi \to i\phi \,, \quad \lambda \to -i\lambda \,. \tag{4.219}$$

The corresponding action is

$$S = \int d^4x \sqrt{-g} \left(\mathscr{R} + \frac{1}{2} \nabla_a \phi \nabla^a \phi - V_0 e^{-\lambda \phi} \right) \qquad (4.220)$$

with a phantom field endowed with the "wrong" sign of the kinetic term. The coupled Einstein-scalar field equations are now

$$G_{ab} = -\nabla_a \phi \nabla_b \phi + \frac{1}{2} g_{ab} \nabla_c \phi \nabla^c \phi - g_{ab} V , \qquad (4.221)$$

$$\Box \phi + \frac{dV}{d\phi} = 0 . \qquad (4.222)$$

The generalized Fonarev metric and phantom scalar are

$$ds^2 = a^2(\eta) \left[-f^2(r) \, d\eta^2 + \frac{dr^2}{f(r)^2} + S^2(r) \, d\Omega_{(2)}^2 \right] , \qquad (4.223)$$

$$\phi(\eta, r) = \frac{1}{\lambda} \ln \left[\frac{V_0 (\lambda^2 + 2)^2}{2 (\lambda^2 + 6)} \right] - \lambda \ln a - \frac{1}{\sqrt{\lambda^2 - 2}} \ln \left(1 - \frac{2w}{r} \right) , \qquad (4.224)$$

where

$$f(r) = \left(1 - \frac{2w}{r} \right)^{\alpha/2} , \qquad \alpha = -\frac{\lambda}{\sqrt{\lambda^2 - 2}} , \qquad (4.225)$$

$$S(r) = r \left(1 - \frac{2w}{r} \right)^{\frac{1-\alpha}{2}} , \qquad a(\eta) = \eta^{-\frac{2}{\lambda^2 + 2}} . \qquad (4.226)$$

Assuming that $\lambda > \sqrt{2}$, it is of interest to understand the physical meaning of the constant w. When $\lambda \gg \sqrt{2}$ it is $a \approx 1$ and $\alpha \approx -1$ and the metric becomes [66]

$$ds^2 \approx -\left(1 - \frac{2w}{r} \right)^{-1} d\eta^2 + \left(1 - \frac{2w}{r} \right) dr^2 + r^2 \left(1 - \frac{2w}{r} \right)^2 d\Omega_{(2)}^2 . \qquad (4.227)$$

Using the coordinate transformation [66]

$$y = r \left(1 - \frac{2w}{r} \right) , \qquad (4.228)$$

the line element (4.227) is rewritten as

$$ds^2 = -\left(1 + \frac{2w}{y} \right) d\eta^2 + \left(1 + \frac{2w}{y} \right)^{-1} dy^2 + y^2 d\Omega_{(2)}^2 , \qquad (4.229)$$

which is the Schwarzschild solution with mass $-w$. Therefore, the parameter w corresponds to the negative mass in this limit and we will we use $-M$ instead of w.

Let us locate the apparent horizons of the generalized Fonarev metric as the parameters M and α vary. By writing the line element as

$$ds^2 = \frac{1}{\eta^{\frac{2\alpha^2-2}{2\alpha^2-1}}} \left[-\left(1 + \frac{2M}{r}\right)^\alpha d\eta^2 + \left(1 + \frac{2M}{r}\right)^{-\alpha} dr^2 \right.$$
$$\left. + r^2 \left(1 + \frac{2M}{r}\right)^{1+\alpha} d\Omega^2_{(2)} \right] \tag{4.230}$$

and replacing the conformal time η with the comoving time t, one obtains

$$ds^2 = -\left(1 + \frac{2M}{r}\right)^\alpha dt^2$$
$$+ a^2(t) \left[\left(1 + \frac{2M}{r}\right)^{-\alpha} dr^2 + r^2 \left(1 + \frac{2M}{r}\right)^{1+\alpha} d\Omega^2_{(2)} \right], \tag{4.231}$$

$$a(t) = (t_0 - t)^{-\frac{2(\alpha^2-1)}{\alpha^2}}, \tag{4.232}$$

where the integration constant t_0 marks the Big Rip and $\alpha < -1$ since $\lambda > \sqrt{2}$. The exponent α is determined by the slope of the potential according to Eq. (4.225). When $M = 0$ the metric (4.231) reduces to a phantom-dominated FLRW one. By setting, for the sake of illustration, $\alpha = -3$ or $\lambda = 3/2$, the line element (4.231) reduces to

$$ds^2 = -\left(1 + \frac{2M}{r}\right)^{-3} dt^2$$
$$+ a^2(t) \left[\left(1 + \frac{2M}{r}\right)^3 dr^2 + r^2 \left(1 + \frac{2M}{r}\right)^{-2} d\Omega^2_{(2)} \right],$$
$$a(t) = (t_0 - t)^{-16/9}. \tag{4.233}$$

In terms of the areal radius $R = ar(1 + 2M/r)^{-1}$, the equation locating the apparent horizons is

$$1 + \frac{8Ma}{R} \left(1 + \sqrt{1 + \frac{8Ma}{R}}\right)^{-1} - \frac{HR}{32} \left(1 + \sqrt{1 + \frac{8Ma}{R}}\right)^5 = 0, \tag{4.234}$$

where $H \equiv \dot{a}/a$ is the Hubble parameter of the FLRW "background". Further setting $x \equiv 1 + \sqrt{1 + \dfrac{8Ma}{R}}$ yields

$$aMHx^4 - 4x^2 + 12x - 8 = 0. \tag{4.235}$$

This quartic equation has only two real positive roots corresponding to a cosmological apparent horizon R_C and a black hole apparent horizon R_{BH} [66]. The qualitative behaviour of the apparent horizons is the same as for the McVittie and generalized McVittie solutions: a black hole apparent horizon inflates while a cosmological apparent horizon shrinks. At a critical time these two apparent horizons coincide and disappear leaving behind a naked singularity [66]. The time reverse of this picture gives the apparent horizons of the Fonarev geometry with canonical scalar.

4.10   Other Analytic Cosmological Black Hole Solutions of the Einstein Equations

Apparently unaware of McVittie's 1933 work, in 1946 Einstein and Straus [44] derived the solution of General Relativity now called Einstein-Straus vacuole or Swiss-cheese model by pasting a Schwarzschild-like region of spacetime onto a dust-dominated FLRW universe across a timelike hypersurface. There is a black hole event horizon in this spacetime and the usual energy conditions are satisfied.

The Einstein-Straus model is discussed at length in the literature (see, e.g., Ref. [92] and references therein) and we will not repeat such discussions here.

The Einstein-Straus vacuole was later generalized to include a cosmological constant, obtaining a Schwarschild-(anti-)de Sitter instead of Schwarzschild interior [7], or to include a fluid with pressure in the interior region [11]. Also the generalization obtained by matching a Schwarschild interior with an inhomogeneous Lemaître-Tolman-Bondi exterior has been studied [12]. The Hawking radiation emitted by the Einstein-Straus black hole has been studied in [148, 149]. It is found that a black hole in an expanding universe is excited to a non-equilibrium state and emits with stronger intensity than a thermal one.

Apparent and trapping horizons were studied also in Oppenheimer-Schneider, Vaidya, and Lemaître-Tolman-Bondi spacetimes [13, 67]. In this last class of models, in particular, the use of different coordinates determines different foliations and, potentially, different apparent horizons. Multiple "S-curve" phenomenology is reported and interpreted as a single apparent horizon tube which goes back and forth in time and, when sliced with hypersurfaces of constant time, produces multiple apparent horizons which appear and disappear in pairs [13].

McClure and Dyer [117] found a spherical solution of the Einstein equations presumably describing a central inhomogeneity in a radiation-dominated universe with a radial heat current, which satisfies the energy conditions everywhere and is perfectly comoving. There is a spacetime singularity at $\bar{r} = m/2$. Similarly, another analytic solution of General Relativity with a dust-dominated "background" universe exhibits singular energy density and a spacetime singularity [117]. Charged versions of the Vaidya, Sultana-Dyer, and Thakurta solutions are reported in Ref. [147].

The Kerr-de Sitter black hole is well known in the literature, but a Kerr-FLRW (or Kerr-McVittie) solution is not reported. Solutions describing spherical shells in

FLRW space were found by studying inflation and phase transitions in the early universe: the most well known are the Coleman-de Luccia [33] and the Farhi-Guth [57] spacetimes.[18]

Common techniques used to generate cosmological black holes consist of:

1. Performing a conformal transformation of a static black hole solution (for example, Schwarzschild in some coordinate system) using a time-dependent conformal factor (usually given by the scale factor of the FLRW "background" universe):

$$g_{ab} \longrightarrow \tilde{g}_{ab} = \Omega^2 g_{ab} = a^2 g_{ab}. \tag{4.236}$$

This technique generates, for example, the Sultana-Dyer black hole and various solutions studied in [116–118].

2. Performing a Kerr-Schild transformation of a static black hole metric g_{ab}:

$$g_{ab} \longrightarrow \bar{g}_{ab} = g_{ab} + \lambda \, k_a k_b, \tag{4.237}$$

where λ is a function and k^c is a null and geodesic vector field with respect to both g_{ab} and $\bar{g}_{ab}$ (see Sect. 1.3.6).

In general, by conformally transforming or Kerr-Schild transforming a "seed" solution of the Einstein equations with standard matter source (including vacuum), it is not guaranteed that the product of this transformation will satisfy the Einstein equations with the same form of matter, or with any reasonable matter source at all. Indeed, one can use the Synge approach consisting of running the Einstein equations from the left to the right, i.e., prescribing a metric motivated in some way and computing the corresponding energy-momentum tensor. But the latter will in general violate the energy conditions and will be physically unreasonable because it is built in a completely artificial way and is devoid of physical content. Indeed, this is the problem of most solutions obtained by conformally or Kerr-Schild transforming a known black hole metric. Moreover, the conformal transformation of a static black hole metric does not always generate a black hole: often it generates a naked singularity instead.

4.11 Conclusions

It is rare to find explicit analytic expressions of the apparent horizons for solutions of the Einstein equations representing cosmological (or other dynamical) black holes. To the best of our knowledge such an expression is available only for the

---

[18]Sometimes one encounters in the literature also dynamical black hole spacetimes which are constructed by hand and are not known to be solutions of the Einstein equations or of the field equations of other theories of gravity (e.g., [16, 60, 105–107]).

extremal charged McVittie black hole and for the "comoving" generalized McVittie spacetime. For other exact solutions of General Relativity the apparent horizon can only be located numerically or given by implicit analytic expressions (for example, for the Schwarzschild-de Sitter-Kottler and the McVittie black holes).

When they represent black holes, the various inhomogeneous solutions considered in this chapter describe:

- Eternal black holes which have not been created in a collapse process but are created together with the universe in the Big Bang, or have existed forever (for example, in a de Sitter background); or
- Black holes that appear when a naked singularity is suddenly covered by an apparent horizon which is created simultaneously with another (cosmological) apparent horizon.

In any case, when a timelike naked singularity is present, the initial value problem [169] is not well posed and the spacetime cannot be obtained as the development of regular Cauchy data.

The subject of dynamical and cosmological black holes is still too young to classify all the possibilities allowed by the Einstein equations in a physically meaningful way, and it is even debatable whether apparent and trapping horizons provide a truly satisfactory notion of dynamical black hole. Nevertheless, one can tentatively group the known solutions of the Einstein equations with these features in two ways:

1. On the basis of the type of matter sourcing the "background" FLRW universe (dust, perfect fluid, imperfect fluid, scalar field, etc.);
2. On the basis of the dynamics and phenomenology of the apparent horizons.

Solutions of General Relativity with a perfect fluid and an electric field include the McVittie and charged McVittie solution (and its special case, the Schwarzschild-de Sitter and Schwarzschild-anti de Sitter black holes); solutions sourced by a canonical scalar field include the Husain-Martinez-Nuñez, the Fonarev, and the phantom Fonarev solutions. At the moment of writing, the phenomenology of apparent horizons distinguishes between the McVittie type with two appearing or disappearing (one black hole and one cosmological) apparent horizons, and the Husain-Martinez-Nuñez-type phenomenology with three apparent horizons. Multiple "S-curve" phenomenology is reported in Lemaître-Tolman-Bondi models [13]. It is not clear whether completely different horizon phenomenologies are possible in Einstein theory.

The rather bizarre phenomenology of the apparent horizons in the solutions examined begs the question of whether they are, after all, physically significant. Naked singularities form during simulations of gravitational collapse but, generally speaking, they are not "typical". General choices of the initial data result in black holes rather than naked singularities. It could well be that the solutions examined here are non-generic or even very special. The "comoving mass solution" is a late-time attractor in the generalized McVittie class but it is not a generic cosmological black hole solution with spherical symmetry [28]. No definitive statement can be made at present.

References

1. Abdalla, E., Afshordi, N., Fontanini, M., Guariento, D.C., Papantonopoulos, E.: Cosmological black holes from self-gravitating fields. Phys. Rev. D **89**, 104018 (2014)
2. Abe, S.: Stability of a collapsed scalar field and cosmic censorship. Phys. Rev. D **38**, 1053 (1988)
3. Afshordi, N., Fontanini, M., Guariento, D.C.: Horndeski meets McVittie: a scalar field theory for accretion onto cosmological black holes. Phys. Rev. D **90**, 084012 (2014)
4. Agnese, A.G., La Camera, M.: Gravitation without black holes. Phys. Rev. D **31**, 1280 (1985)
5. Amendola, L., Tsujikawa, S.: Dark Energy, Theory and Observations. Cambridge University Press, Cambridge (2010)
6. Babichev, E., Dokuchaev, V., Eroshenko, Yu.: Black hole mass decreasing due to phantom energy accretion. Phys. Rev. Lett. **93**, 021102 (2004)
7. Balbinot, R., Bergamini, R., Comastri, A.: Phys. Rev. D **38**, 2415 (1988)
8. Barris, B., et al.: Twenty-three high-redshift supernovae from the Institute for Astronomy Deep Survey: doubling the supernova sample at $z > 0.7$. Astrophys. J. **602**, 571 (2004)
9. Bergman, O., Leipnik, R.: Phys. Rev. **107**, 1157 (1957)
10. Bochicchio, I., Faraoni, V.: A Lemaître-Tolman-Bondi cosmological wormhole. Phys. Rev. D **82**, 044040 (2010)
11. Bona, C., Stela, J.: "Swiss cheese" models with pressure. Phys. Rev. D **36**, 2915 (1987)
12. Bonnor, W.B.: A generalization of the Einstein-Straus vacuole. Class. Quantum Grav. **17**, 2739 (2000)
13. Booth, I., Brits, L., Gonzalez, J.A., Van den Broeck, V.: Marginally trapped tubes and dynamical horizons. Class. Quantum Grav. **23**, 413 (2006)
14. Bousso, R.: Adventures in de Sitter space. Preprint arXiv:hep-th/0205177
15. Brevik, I., Nojiri, S., Odintsov, S.D., Vanzo, L.: Entropy and universality of the Cardy-Verlinde formula in a dark energy universe. Phys. Rev. D **70**, 043520 (2004)
16. Brown, B.A., Lindesay, J.: Class. Quantum Grav. **26**, 045010 (2009)
17. Brown, I., Behrend, J., Malik, K.: Gauges and cosmological backreaction. J. Cosmol. Astropart. Phys. **11**, 027 (2009)
18. Buchdahl, H.A.: Static solutions of the brans-dicke equations. Int. J. Theor. Phys. **6**, 407 (1972)
19. Buchdahl, H.A.: Isotropic coordinates and Schwarzschild metric. Int. J. Theor. Phys. **24**, 731 (1985)
20. Buchert, T.: On average properties of inhomogeneous fluids in general relativity: dust cosmologies. Gen. Rel. Gravit. **32**, 105 (2000)
21. Buchert, T.: Backreaction issues in relativistic cosmology and the dark energy debate. AIP Conf. Proc. **910**, 361 (2007)
22. Buchert, T.: Gen. Rel. Gravit. **40**, 467 (2008)
23. Buchert, T., Carfora, M.: Regional averaging and scaling in relativistic cosmology. Class. Quantum Grav. **19**, 6109 (2002)
24. Buchert, T., Carfora, M.: On the curvature of the present-day universe. Class. Quantum Grav. **25**, 195001 (2008)
25. Caldwell, R.R., Kamionkowski, M., Weinberg, N.N.: Phantom energy and cosmic doomsday. Phys. Rev. Lett. **91**, 071301 (2003)

26. Capozziello, S., Carloni, S., Troisi, A.: Quintessence without scalar fields. Recent Res. Dev. Astron. Astrophys. **1**, 625 (2003)
27. Carr, B.J.: Primordial black holes: do they exist and are they useful? Preprint astro-ph/0511743
28. Carrera, M., Giulini, D.: Influence of global cosmological expansion on local dynamics and kinematics. Rev. Mod. Phys. **82**, 169 (2010)
29. Carrera, M., Giulini, D.: On the generalization of McVittie's model for an inhomogeneity in a cosmological spacetime. Phys. Rev. D **81**, 043521 (2010)
30. Carroll, S.M., Duvvuri, V., Trodden, M., Turner, M.S.: Is cosmic speed-up due to new gravitational physics? Phys. Rev. D **70**, 043528 (2004)
31. Castro, A., Rodriguez, M.J.: Universal properties and the first law of black hole inner mechanics. Phys. Rev. D **86**, 024008 (2012)
32. Chen, S., Jing, J.: Quasinormal modes of a black hole surrounded by quintessence. Class. Quantum Grav. **22**, 4651 (2005)
33. Coleman, S.R., De Luccia, F.: Gravitational effects on and of vacuum decay. Phys. Rev. D **21**, 3305 (1980)
34. Coley, A.A., van den Hoogen, R.J.: Dynamics of multi-scalar-field cosmological models and assisted inflation. Phys. Rev. D **62**, 023517 (2000)
35. Cvetic, M., Larsen, F.: General rotating black holes in string theory: greybody factors and event horizons. Phys. Rev. D **56**, 4994 (1997)
36. Cvetic, M., Larsen, F.: Greybody factors and charges in Kerr/CFT. J. High Energy Phys. **0909**, 088 (2009)
37. Cvetic, M., Gibbons, G.W., Pope, C.N.: Universal area product formulae for rotating and charged black holes in four and higher dimensions. Phys. Rev. Lett. **106**, 121301 (2011)
38. da Silva, A., Fontanini, M., Guariento, D.C.: How the expansion of the universe determines the causal structure of McVittie spacetimes. Phys. Rev. D **87**, 064030 (2013)
39. da Silva, A., Guariento, D.C., Molina, C.: Cosmological black holes and white holes with time-dependent mass. Phys. Rev. D **91**, 084043 (2015)
40. De Felice, A., Tsujikawa, S.: $f(R)$ theories. Living Rev. Relat. **13**, 3 (2010)
41. de Freitas Pacheco, J.A., Horvath, J.E.: Generalized second law and phantom cosmology. Class. Quantum Grav. **24**, 5427 (2007)
42. Dicke, R.H.: Mach's principle and invariance under transformation of units. Phys. Rev. **125**, 2163 (1962)
43. Di Criscienzo, R., Nadalini, M., Vanzo, L., Zerbini, S., Zoccatelli, G.: On the Hawking radiation as tunneling for a class of dynamical black holes. Phys. Lett. B **657**, 107 (2007)
44. Einstein, A., Straus, E.G.: The influence of the expansion of space on the gravitation fields surrounding the individual stars. Rev. Mod. Phys. **17**, 120 (1945)
45. Einstein, A., Straus, E.G.: Corrections and additional remarks to our paper: the influence of the expansion of space on the gravitation fields surrounding the individual stars. Rev. Mod. Phys. **18**, 148 (1946)
46. Faraoni, V.: Cosmology in Scalar-Tensor Gravity. Kluwer Academic, Dordrecht (2004)
47. Faraoni, V.: Hawking temperature of expanding cosmological black holes. Phys. Rev. D **76**, 104042 (2007)
48. Faraoni, V.: Analysis of the Sultana-Dyer cosmological black hole solution of the Einstein equations. Phys. Rev. D **80**, 044013 (2009)
49. Faraoni, V.: Evolving black hole horizons in general relativity and alternative gravity. Galaxies **1**, 114 (2013)
50. Faraoni, V., Israel, W.: Dark energy, wormholes, and the big rip. Phys. Rev. D **71**, 064017 (2005)
51. Faraoni, V., Jacques, A.: Cosmological expansion and local physics. Phys. Rev. D **76**, 063510 (2007)
52. Faraoni, V., Vitagliano, V.: Horizon thermodynamics and spacetime mappings. Phys. Rev. D **89**, 064015 (2014)

53. Faraoni, V., Zambrano Moreno, A.F.: Are quantization rules for horizon areas universal? Phys. Rev. D **88**, 044011 (2013)
54. Faraoni, V., Gao, C., Chen, X., Shen, Y.-G.: What is the fate of a black hole embedded in an expanding universe? Phys. Lett. B **671**, 7 (2009)
55. Faraoni, V., Zambrano Moreno, A.F., Nandra, R.: Making sense of the bizarre behavior of horizons in the McVittie spacetime. Phys. Rev. D **85**, 083526 (2012)
56. Faraoni, V., Zambrano Moreno, A.F., Prain, A.: Charged McVittie spacetime. Phys. Rev. D **89**, 103514 (2013)
57. Farhi, E., Guth, A.H., Guven, J.: Is it possible to create a universe in the laboratory by quantum tunneling? Nucl. Phys. B **339**, 417 (1990)
58. Ferraris, M., Francaviglia, M., Spallicci, A.: Associated radius, energy and pressure of McVittie's metric in its astrophysical application. Nuovo Cimento **111B**, 1031 (1996)
59. Figueras, P., Hubeny, V.E., Rangamani, M., Ross, S.F.: Dynamical black holes and expanding plasmas. J. High Energy Phys. **0904**, 137 (2009)
60. Finch, T.K., Lindesay, J.: Global causal structure of a transient black object. Preprint arXiv:1110.6928
61. Fisher, I.Z.: Scalar mesostatic field with regard for gravitational effects. Zh. Eksp. Teor. Fiz. **18**, 636 (1948) (translated in arXiv:gr-qc/9911008)
62. Fonarev, O.A.: Exact Einstein scalar field solutions for formation of black holes in a cosmological setting. Class. Quantum Grav. **12**, 1739 (1995)
63. Galli, P., Ortin, T., Perz, J., Shahbazi, C.S.: Non-extremal black holes of $N = 2, d = 4$ supergravity. J. High Energy Phys. **1107**, 041 (2011)
64. Gao, C.J., Zhang, S.N.: Reissner-Nordstrom metric in the Friedman-Robertson-Walker universe. Phys. Lett. B **595**, 28 (2004)
65. Gao, C.J., Zhang, S.N.: Higher dimensional Reissner-Nordstrom-FRW metric. Gen. Rel. Gravit. **38**, 23 (2006)
66. Gao, C., Chen, X., Faraoni, V., Shen, Y.-G.: Does the mass of a black hole decrease due to accretion of phantom energy? Phys. Rev. D **78**, 024008 (2008)
67. Gao, C., Chen, X., Shen, Y.-G., Faraoni, V.: Black holes in the universe: generalized Lemaître-Tolman-Bondi solutions. Phys. Rev. D **84**, 104047 (2011)
68. Gomes, H., Gryb, S., Koslowski, T.: Einstein gravity as a 3D conformally invariant theory. Class. Quantum Grav. **28**, 045005 (2011)
69. Gonzalez, J.A., Guzman, F.S.: Accretion of phantom scalar field into a black hole. Phys. Rev. D **79**, 121501 (2009)
70. Gonzalez-Diaz, P.F., Siguenza, C.L.: Phantom thermodynamics. Nucl. Phys. B **697**, 363 (2004)
71. Green, S.R., Wald, R.M.: New framework for analyzing the effects of small scale inhomogeneities in cosmology. Phys. Rev. D **83**, 084020 (2011)
72. Guariento, D.C., Horvath, J.E., Custodio, P.S., de Freitas Pacheco, J.A.: Evolution of primordial black holes in a radiation and phantom energy environment. Gen. Rel. Gravit. **40**, 1593 (2008)
73. Guo, Z.-K., Piao, Y.-S., Cai, R.-G., Zhang, Y.-Z.: Cosmological scaling solutions and cross coupling exponential potential. Phys. Lett. B **576**, 12 (2003)
74. Harada, T., Carr, B.J.: Upper limits on the size of a primordial black hole. Phys. Rev. D **71**, 104009 (2005)
75. Harada, T., Carr, B.J.: Growth of primordial black holes in a universe containing a massless scalar field. Phys. Rev. D **71**, 104010 (2005)
76. Harada, T., Maeda, H., Carr, B.J.: Nonexistence of self-similar solutions containing a black hole in a universe with a stiff fluid or scalar field or quintessence. Phys. Rev. D **74**, 024024 (2006)
77. Hawking, S.W., Ellis, G.F.R.: The Large Scale Structure of Space-Time. Cambridge University Press, Cambridge (1973)
78. He, X., Wang, B., Wu, S.-F., Lin, C.-Y.: Quasinormal modes of black holes absorbing dark energy. Phys. Lett. B **673**, 156 (2009)

79. Horowitz, G.T., Maldacena, J.M., Strominger, A.: Nonextremal black hole microstates and U duality. Phys. Lett. B **383**, 151 (1996)
80. Hsu, D.H., Jenskins, A., Wise, M.B.: Gradient instability for $w < -1$. Phys. Lett. B **597**, 270 (2004)
81. Hubeny, V.: The fluid/gravity correspondence: a new perspective on the membrane paradigm. Class. Quantum Grav. **28**, 114007 (2011)
82. Husain, V., Martinez, E.A., Nuñez, D.: Exact solution for scalar field collapse. Phys. Rev. D **50**, 3783 (1994)
83. Izquierdo, G., Pavon, D.: The Generalized second law in phantom dominated universes in the presence of black holes. Phys. Lett. B **639**, 1 (2006)
84. Janis, A.I., Newman, E.T., Winicour, J.: Reality of the Schwarzschild singularity. Phys. Rev. Lett. **20**, 878 (1968)
85. Kaloper, N., Kleban, M., Martin, D.: McVittie's legacy: black holes in an expanding universe. Phys. Rev. D **81**, 104044 (2010)
86. Kitada, Y., Maeda, K.: Cosmic no hair theorem in homogeneous space-times. 1. Bianchi models. Class. Quantum Grav. **10**, 703 (1993)
87. Knop, R., et al.: New constraints on Ω_M, Ω_A, and w from an independent set of 11 high-redshift supernovae observed with the Hubble Space Telescope. Astrophys. J. **598**, 102 (2003)
88. Kolb, E.W., Matarrese, S., Riotto, A.: On cosmic acceleration without dark energy. New J. Phys. **8**, 322 (2006)
89. Kolb, E., Marra, V., Matarrese, S.: Cosmological background solutions and cosmological backreactions. Gen. Rel. Gravit. **42**, 1399 (2010)
90. Komatsu, E., et al.: Seven-year Wilkinson Microwave Anisotropy Probe (WMAP*) observations: cosmological interpretation. Astrophys. J. (Suppl.) **192**, 18 (2011)
91. Kottler, F.: Über die physikalischen ndlagen der Einsteinschen gravitationstheorie. Ann. Phys. (Leipzig) **361**, 401 (1918)
92. Krasiński, A.: Inhomogeneous Cosmological Models. Cambridge University Press, Cambridge (1997)
93. Kustaanheimo, P., Qvist, B.: A note on some general solutions of the Einstein field equations in a spherically symmetric world. Comm. Phys. Math. Soc. Sci. Fennica **13**(16), 1 (1948) (reprinted in Gen. Rel. Gravit. **30**, 659 (1998))
94. Lake, K., Abdelqader, M.: More on McVittie's legacy: a Schwarzschild-de Sitter black and white hole embedded in an asymptotically ΛCDM cosmology. Phys. Rev. D **84**, 044045 (2011)
95. Landry, P., Abdelqader, M., Lake, K.: McVittie solution with a negative cosmological constant. Phys. Rev. D **86**, 084002 (2012)
96. Larena, J.: Spatially averaged cosmology in an arbitrary coordinate system. Phys. Rev. D **79**, 084006 (2009)
97. Larena, J., Buchert, T., Alimi, J.-M.: Correspondence between kinematical backreaction and scalar field cosmologies: the 'morphon field'. Class. Quantum Grav. **23**, 6379 (2006)
98. Larena, J., Alimi, J.-M., Buchert, T., Kunz, M., Corasaniti, P.: Testing backreaction effects with observations. Phys. Rev. D **79**, 083011 (2009)
99. Larsen, F.: String model of black hole microstates. Phys. Rev. D **56**, 1005 (1997)
100. Le Delliou, M., Mimoso, J.P., Mena, F.C., Fontanini, M., Guariento, D.C., Abdalla, E.: Separating expansion and collapse in general fluid models with heat flux. Phys. Rev. D **88**, 027301 (2013)
101. Li, N., Schwarz, D.J.: Onset of cosmological backreaction. Phys. Rev. D **76**, 083011 (2007)
102. Li, N., Schwarz, D.J.: Scale dependence of cosmological backreaction. Phys. Rev. D **78**, 083531 (2008)
103. Liddle, A.R., Mazumdar, A., Schunck, F.E.: Assisted inflation. Phys. Rev. D **58**, 061301 (1998)
104. Lima, J.A.S., Alcaniz, J.S.: Thermodynamics and spectral distribution of dark energy. Phys. Lett. B **600**, 191 (2004)
105. Lindesay, J.: Found. Phys. **37**, 1181 (2007)

106. Lindesay, J.: Foundations of Quantum Gravity, p. 282. Cambridge University Press, Cambridge (2013)
107. Lindesay, J., Sheldon, P.: Class. Quantum Grav. **27**, 215015 (2010)
108. Maeda, H., Harada, T., Carr, B.J.: Self-similar cosmological solutions with dark energy. II. Black holes, naked singularities, and wormholes. Phys. Rev. D **77**, 024023 (2008)
109. Maeda, H., Harada, T., Carr, B.J.: Cosmological wormholes. Phys. Rev. D **79**, 044034 (2009)
110. Majhi, B.R.: Thermodynamics of Sultana-Dyer black hole. J. Cosmol. Astropart. Phys. **1405**, 014 (2014)
111. Malik, K.A., Wands, D.: Dynamics of assisted inflation. Phys. Rev. D **59**, 123501 (1999)
112. Marra, V.: A back-reaction approach to dark energy. Preprint arXiv:0803.3152
113. Marra, V., Kolb, E., Matarrese, S., Riotto, A.: Cosmological observables in a Swiss-cheese universe. Phys. Rev. D **76**, 123004 (2007)
114. Marra, V., Kolb, E., Matarrese, S.: Light-cone averages in a Swiss-cheese universe. Phys. Rev. D **77**, 023003 (2008)
115. Mashhoon, B., Partovi, M.H.: Gravitational collapse of a charged fluid sphere. Phys. Rev. D **20**, 2455 (1979)
116. McClure, M.L.: Cosmological black holes as models of cosmological inhomogeneities. PhD thesis, University of Toronto (2006)
117. McClure, M.L., Dyer, C.C.: Asymptotically Einstein-de Sitter cosmological black holes and the problem of energy conditions. Class. Quantum Grav. **23**, 1971 (2006)
118. McClure, M.L., Dyer, C.C.: Matching radiation-dominated and matter-dominated Einstein-de Sitter universes and an application for primordial black holes in evolving cosmological backgrounds. Gen. Rel. Gravit. **38**, 1347 (2006)
119. McVittie, G.C.: The mass-particle in an expanding universe. Mon. Not. R. Astr. Soc. **93**, 325 (1933)
120. Meessen, P., Ortin, T., Perz, J., Shahbazi, C.S.: Black holes and black strings of $N = 2, d = 5$ supergravity in the H-FGK formalism. J. High Energy Phys. **1209**, 001 (2012)
121. Miller, J.C., Musco, I.: Causal horizons and topics in structure formation. Preprint arXiv:1412.8660
122. Mosheni Sadjadi, H.: Generalized second law in a phantom-dominated universe. Phys. Rev. D **73**, 0635325 (2006)
123. Nandra, R., Lasenby, A.N., Hobson, M.P.: The effect of a massive object on an expanding universe. Mon. Not. R. Astr. Soc. **422**, 2931 (2012)
124. Nandra, R., Lasenby, A.N., Hobson, M.P.: The effect of an expanding universe on massive objects. Mon. Not. R. Astr. Soc. **422**, 2945 (2012)
125. Nariai, H.: On some static solutions of Einstein's gravitational field equations in a spherically symmetric case. Sci. Rep. Tohoku Univ. **34**, 160 (1950)
126. Nariai, H.: On a new cosmological solution of Einstein's field equations of gravitation. Sci. Rep. Tohoku Univ. **35**, 62 (1951)
127. Nojiri, S., Odintsov, S.D.: Final state and thermodynamics of a dark energy universe. Phys. Rev. D **70**, 103522 (2004)
128. Nojiri, S., Odintsov, S.D.: Quantum escape of sudden future singularity. Phys. Lett. B **595**, 1 (2004)
129. Nolan, B.C.: Sources for McVittie's mass particle in an expanding universe. J. Math. Phys. **34**, 1 (1993)
130. Nolan, B.C.: A Point mass in an isotropic universe: existence, uniqueness and basic properties. Phys. Rev. D **58**, 064006 (1998)
131. Nolan, B.C.: A Point mass in an isotropic universe. 2. Global properties. Class. Quantum Grav. **16**, 1227 (1999)
132. Nolan, B.C.: A Point mass in an isotropic universe 3. The region $R \leq 2m$. Class. Quantum Grav. **16**, 3183 (1999)
133. Paranjape, A., Singh, T.P.: The possibility of cosmic acceleration via spatial averaging in Lemaitre-Tolman-Bondi models. Class. Quantum Grav. **23**, 6955 (2006)

134. Paranjape, A., Singh, T.P.: The spatial averaging limit of covariant macroscopic gravity: scalar corrections to the cosmological equations. Phys. Rev. D **76**, 044006 (2007)
135. Paranjape, A., Singh, T.P.: Explicit cosmological coarse graining via spatial averaging. Gen. Rel. Gravit. **40**, 139 (2008)
136. Perlmutter, S., et al.: Discovery of a supernova explosion at half the age of the Universe. Nature **391**, 51 (1998)
137. Perlmutter, S., et al.: Measurements of Ω and Λ from 42 high-redshift supernovae. Astrophys. J. **517**, 565 (1999)
138. Poisson, E., Israel, W.: The internal structure of black holes. Phys. Rev. D **41**, 1796 (1990)
139. Räsänen, S.: Dark energy from backreaction. J. Cosmol. Astropart. Phys. **02**, 003 (2004)
140. Räsänen, S.: Backreaction in the Lemaitre-Tolman-Bondi model. J. Cosmol. Astropart. Phys. **11**, 010 (2004)
141. Raychaudhuri, A.K.: Theoretical Cosmology. Clarendon Press, Oxford (1979)
142. Riess, A.G., et al.: Observational evidence from supernovae for an accelerating universe and a cosmological constant. Astron. J. **116**, 1009 (1998)
143. Riess, A.G., et al.: An indication of evolution of type Ia supernovae from their risetimes. Astron. J. **118**, 2668 (1999)
144. Riess, A.G., et al.: The farthest known supernova: support for an accelerating universe and a glimpse of the epoch of deceleration. Astrophys. J. **560**, 49 (2001)
145. Riess, A.G., et al.: Type Ia supernova discoveries at $z > 1$ from the Hubble Space Telescope: evidence for past deceleration and constraints on dark energy evolution. Astron. J. **607**, 665 (2004)
146. Roberts, M.D.: Massless scalar static spheres. Astrophys. Space Sci. **200**, 331 (1993)
147. Rodrigues, M.G., Zanchin, V.T.: Charged black holes in expanding Einstein-de Sitter universes. Class. Quantum Grav. **32**, 115004 (2015)
148. Saida, H.: Hawking radiation in the Swiss-cheese universe. Class. Quantum Grav. **19**, 3179 (2002)
149. Saida, H., Harada, T., Maeda, H.: Black hole evaporation in an expanding universe. Class. Quantum Grav. **24**, 4711 (2007)
150. Shah, Y.P., Vaidya, P.C.: Gravitational field of a charged particle embedded in a homogeneous universe. Tensor **19**, 191 (1968)
151. Shankaranarayanan, S.: Temperature and entropy of Schwarzschild.de Sitter space-time. Phys. Rev. D **67**, 084026 (2003)
152. Sotiriou, T.P.: $f(R)$ gravity and scalar-tensor theory. Class. Quantum Grav. **23**, 5117 (2006)
153. Sotiriou, T.P.: PhD thesis, International School for Advanced Studies, Trieste (2007) (preprint arXiv:0710.4438)
154. Sotiriou, T.P.: In: Kleinert, H., Jantzen, R.T., Ruffini, R. (eds.) Proceedings of the Eleventh Marcel Grossmann Meeting on General Relativity, pp. 1223–1226. World Scientific, Singapore (2008) (preprint arXiv:gr-qc/0611158)
155. Sotiriou, T.P., Faraoni, V.: $f(R)$ theories of gravity. Rev. Mod. Phys. **82**, 451 (2010)
156. Sotiriou, T.P., Liberati, S.: Metric-affine $f(R)$ theories of gravity. Ann. Phys. (N.Y.) **322**, 935 (2007)
157. Sotiriou, T.P., Liberati, S.: The metric-affine formalism of $f(R)$ gravity. J. Phys. Conf. Ser. **68**, 012022 (2007)
158. Sultana, J., Dyer, C.C.: Cosmological black holes: a black hole in the Einstein-de Sitter universe. Gen. Rel. Gravit. **37**, 1349 (2005)
159. Sun, C.-Y.: Phantom energy accretion onto black holes in a cyclic universe. Phys. Rev. D **78**, 064060 (2008)
160. Sushkov, S.V., Kim, S.-W.: Cosmological evolution of a ghost scalar field. Gen. Rel. Gravit. **36**, 1671 (2004)
161. Sussman, R.: Conformal structure of a Schwarzschild black hole immersed in a Friedman universe. Gen. Rel. Gravit. **17**, 251 (1985)
162. Tonry, J.L., et al.: Cosmological results from high-z supernovae. Astrophys. J. **594**, 1 (2003)

163. Tsagas, C.G., Challinor, A., Maartens, R.: Relativistic cosmology and large-scale structure. Phys. Rep. **465**, 61 (2008)
164. Virbhadra, K.S.: Janis-Newman-Winicour and Wyman solutions are the same. Int. J. Mod. Phys. A **12**, 4831 (1997)
165. Visser, M.: Quantization of area for event and Cauchy horizons of the Kerr-Newman black hole. J. High Energy Phys. **1206**, 023 (2012)
166. Visser, M.: Area products for stationary black hole horizons. Phys. Rev. D **88**, 044014 (2013)
167. Vitagliano, V., Liberati, S., Faraoni, V.: Averaging inhomogeneities in scalar-tensor cosmology. Class. Quantum Grav. **26**, 215005 (2009)
168. Vollick, D.N.: Phys. Rev. D **68**, 063510 (2003)
169. Wald, R.M.: General Relativity. Chicago University Press, Chicago (1984)
170. Weyl, H.: Zur Gravitationstheorie. Ann. Phys. (Leipzig) **54**, 117 (1917)
171. Weyl, H.: Gravitation und Elektrizit. Stz. Preuss. Akad. Wiss. **1**, 465 (1918)
172. Weyl, H.: Space, Time, Matter. Dover, New York (1950)
173. Wiltshire, D.L.: Cosmic clocks, cosmic variance and cosmic averages. New J. Phys. **9**, 377 (2007)
174. Wiltshire, D.L.: Exact solution to the averaging problem in cosmology. Phys. Rev. Lett. **99**, 251101 (2007)
175. Wyman, M.: Static spherically symmetric scalar fields in general relativity. Phys. Rev. D **24**, 839 (1981)

Chapter 5
Cosmological Inhomogeneities in Alternative Gravity

> To such an extent does nature delight and abound in variety that
> among her trees there is not one plant to be found which is
> exactly like another; and not only among the plants, but among
> the boughs, the leaves and the fruits, you will not find one which
> is exactly similar to another.
>
> —Leonardo da Vinci

5.1 Introduction

After studying inhomogeneous spacetimes describing central condensations embedded in FLRW spaces in Einstein's theory, we now turn to qualitatively similar spacetimes in the context of alternative theories of gravity. In addition to the reasons already mentioned in the previous chapter for the study of cosmological black holes, there is motivation to extend the analysis to alternative theories of gravity. We have already mentioned that the McVittie spacetime is a solution of the cuscuton theory and that the generalized McVittie solution solves the field equations of Horndeski gravity. Let us consider now, in particular, Brans-Dicke [5] and scalar-tensor theories of gravity [4, 40, 56]: these theories are the prototypical alternatives to General Relativity and allow for a varying gravitational strength. While usually only the cosmological variation of the gravitational "constant" G_{eff} with time is studied, inhomogeneous spacetimes describing central objects embedded in FLRW "backgrounds" constitute toy models to study the *spatial* variation of G_{eff} [13, 48]. Another motivation is that explicit analytic examples of evolving apparent horizons would be useful to study Hawking radiation and black hole thermodynamics in a fully dynamical situation in alternative gravity, and bring into light possible differences with General Relativity. Although we restrict ourselves to 4-dimensional spacetime manifolds, once one begins to study alternative theories the door is open for further generalization. For example, exact cosmological and time-dependent black holes are of interest in higher-dimensional Gauss-Bonnet gravity [41], and other examples arise from intersecting branes in supergravity [36].

The known solutions of field equations alternative to the Einstein equations with the desired properties are usually spherically symmetric. It is unknown whether, in general, the concepts of Misner-Sharp-Hernandez mass and Kodama vector can be

© Springer International Publishing Switzerland 2015

V. Faraoni, *Cosmological and Black Hole Apparent Horizons*,
Lecture Notes in Physics 907, DOI 10.1007/978-3-319-19240-6_5

extended beyond General Relativity,[1] and some of the quantities needed to write down the first law of thermodynamics become obscure. Since the Misner-Sharp-Hernandez mass is a special case of the Hawking-Hayward quasi-local energy [31], it seems rather natural to use this construct as the internal energy when abandoning spherical symmetry. The Kodama vector, however, is not defined once this special symmetry is given up.

The concepts of apparent and trapping horizon do not depend on the field equations of the theory. Moreover, there have been several studies of horizon entropy for event horizons in alternative gravities (see Ref. [22] for a review).

5.2 Brans-Dicke Cosmological Black Holes

In string theories [30] a dilaton field coupled non-minimally to the Ricci curvature mimics a Brans-Dicke scalar field (in its low-energy limit, bosonic string theory reduces precisely to an $\omega_0 = -1$ Brans-Dicke theory [6, 28]). This fact has added motivation for the study of scalar-tensor theories in general. Brans-Dicke gravity corresponds to the action [5]

$$S_{\mathrm{BD}} = \int d^4x \sqrt{-g} \left[\phi \mathscr{R} - \frac{\omega_0}{\phi} g^{ab} \nabla_a \phi \nabla_b \phi + 2\kappa \mathscr{L}^{(\mathrm{m})} \right] , \qquad (5.1)$$

where $\kappa \equiv 8\pi$, $\mathscr{L}^{(\mathrm{m})}$ is the matter Lagrangian density, and the Brans-Dicke scalar field ϕ effectively plays the role of the inverse of the gravitational coupling. ω_0 is the "Brans-Dicke parameter".

Contrary to General Relativity, even static, asymptotically flat, spherically symmetric black holes in scalar-tensor gravity are not forced to be Schwarzschild: the Jebsen-Birkhoff theorem is peculiar to Einstein theory and breaks down in more general contexts, even in the simplest Brans-Dicke case (5.1). In this theory, what can be rescued is only a very weak form of the theorem: if the Brans-Dicke scalar field is required to be time-independent in electrovacuo, then the metric is static (but not necessarily the Schwarzschild or Reissner-Nordström one) [23]. In this form, however, the theorem is not very useful and it allows for substantial departures from the Schwarzschild geometry.[2] To the extent that astrophysical black holes can be considered as isolated, however, all physically reasonable (that is, stable and not fine-tuned) black holes of scalar-tensor gravity reduce to general-relativistic black holes [51]. Let us consider now asymptotically FLRW solutions of Brans-Dicke theory which do not fall into this category. The best known solutions are the analytic ones found by Clifton, Mota and Barrow [13] (discussed in Ref. [27], which we follow here) and the numerical ones of Sakai and Barrow [48]. We focus on the former.

---

[1] See Ref. [14] for a proposal.

[2] The Jebsen-Birkhoff theorem, however, holds in Gauss-Bonnet gravity [58].

5.2.1 Clifton-Mota-Barrow Black Holes

The matter source is assumed to be a perfect fluid with energy density $\rho^{(m)}$, pressure $P^{(m)}$, and equation of state $P^{(m)} = (\gamma - 1)\,\rho^{(m)}$, with $\gamma = $ constant [13]. The spherically symmetric and dynamical *Clifton-Mota-Barrow line element* is [13]

$$ds^2 = -e^{\nu(\bar{r})} dt^2 + a^2(t) e^{\mu(\bar{r})} (d\bar{r}^2 + \bar{r}^2 d\Omega_{(2)}^2), \tag{5.2}$$

where

$$e^{\nu(\bar{r})} = \left(\frac{1 - \frac{m}{2\alpha\bar{r}}}{1 + \frac{m}{2\alpha\bar{r}}} \right)^{2\alpha} \equiv A^{2\alpha}, \tag{5.3}$$

$$e^{\mu(\bar{r})} = \left(1 + \frac{m}{2\alpha\bar{r}} \right)^4 A^{\frac{2}{\alpha}(\alpha-1)(\alpha+2)}, \tag{5.4}$$

$$a(t) = a_0 \left(\frac{t}{t_0} \right)^{\frac{2\omega_0(2-\gamma)+2}{3\omega_0\gamma(2-\gamma)+4}} \equiv a_* t^\beta, \tag{5.5}$$

$$\phi(t, \bar{r}) = \phi_0 \left(\frac{t}{t_0} \right)^{\frac{2(4-3\gamma)}{3\omega_0\gamma(2-\gamma)+4}} A^{-\frac{2}{\alpha}(\alpha^2-1)}, \tag{5.6}$$

$$\alpha = \sqrt{\frac{2(\omega_0 + 2)}{2\omega_0 + 3}}, \tag{5.7}$$

$$\rho^{(m)}(t, \bar{r}) = \rho_0^{(m)} \left(\frac{a_0}{a(t)} \right)^{3\gamma} A^{-2\alpha}, \tag{5.8}$$

ω_0 is the Brans-Dicke parameter, m is a mass parameter, $\alpha, \phi_0, a_0, \rho_0^{(m)}$ and t_0 are positive constants ($\phi_0, \rho_0^{(m)}$, and t_0 are not independent). $\bar{r}$ is the isotropic radius related to the Schwarzschild radial coordinate r by

$$r \equiv \bar{r} \left(1 + \frac{m}{2\alpha\bar{r}} \right)^2, \tag{5.9}$$

so that

$$dr = \left(1 - \frac{m^2}{4\alpha^2\bar{r}^2} \right) d\bar{r}. \tag{5.10}$$

The constant α is real for $\omega_0 < -2$ and for $\omega_0 > -3/2$. As customary in Brans-Dicke theory [5, 17, 29], we assume that $\omega_0 > -3/2$ and $\beta \geq 0$. The metric (5.2) is separable and reduces to the spatially flat FLRW one if m is set to zero. If $\gamma \neq 2$, setting $\omega_0 = (\gamma - 2)^{-1}$ yields $\beta = 0$ and the geometry becomes static (interestingly, the scalar field remains time-dependent). If instead $\gamma = 2$ or $\gamma = 4/3$, then $\beta = 1/2$

and the scale factor $a(t) \sim \sqrt{t}$ irrespective of the value of the parameter ω_0. These special cases will be discussed separately.

The areal radius is

$$R = a(t)\bar{r}\left(1 + \frac{m}{2\alpha\bar{r}}\right)^2 A^{\frac{1}{\alpha}(\alpha-1)(\alpha+2)}$$

$$= a(t)rA^{\frac{1}{\alpha}(\alpha-1)(\alpha+2)} \tag{5.11}$$

and the line element is written as

$$ds^2 = -A^{2\alpha}dt^2 + a^2(t)A^{\frac{2}{\alpha}(\alpha^2-2)}dr^2 + R^2 d\Omega_{(2)}^2 . \tag{5.12}$$

Using the relation between differentials

$$dr = \frac{dR - \dot{a}(t)rA^{\frac{1}{\alpha}(\alpha-1)(\alpha+2)}dt}{a(t)A^{\frac{1}{\alpha}(\alpha-1)(\alpha+2)-2}\left[A^2 + \frac{m}{\alpha^2 r}(\alpha-1)(\alpha+2)\right]}, \tag{5.13}$$

the line element becomes

$$ds^2 = -\left[A^{2\alpha} - \frac{\dot{a}^2(t)r^2}{B^2(\bar{r})}A^{\frac{2}{\alpha}(\alpha^2+2\alpha-2)}\right]dt^2$$

$$-2\frac{\dot{a}(t)r}{B^2(\bar{r})}A^{\frac{\alpha^2+3\alpha-2}{\alpha}}dRdt + \frac{A^2(\bar{r})}{B^2(\bar{r})}dR^2 + R^2 d\Omega_{(2)}^2 , \tag{5.14}$$

where

$$B(\bar{r}) \equiv A^2(\bar{r}) + \frac{(\alpha-1)(\alpha+2)}{\alpha^2}\frac{m}{r} \tag{5.15}$$

is positive because $\alpha = \sqrt{\dfrac{2(\omega_0 + 2)}{2\omega_0 + 3}} \geq 1$.

The $dtdR$ cross-term is eliminated by introducing the new time $\bar{t}$ defined by [27]

$$d\bar{t} = \frac{1}{F(t, R)}\left[dt + \psi(t, R)dr\right] , \tag{5.16}$$

where $\psi(t, R)$ is a function to be determined and $F(t, R)$ is an integrating factor, as usual. With this coordinate, the line element becomes

$$ds^2 = -\left[A^{2\alpha} - \frac{\dot{a}^2(t)r^2}{B(\bar{r})^2}A^{\frac{2}{\alpha}(\alpha^2+2\alpha-2)}\right]F^2 d\bar{t}^2$$

$$+ \left\{2\psi F\left[A^{2\alpha} - \frac{\dot{a}^2(t)r^2}{B(\bar{r})^2}A^{\frac{2}{\alpha}(\alpha^2+2\alpha-2)}\right] - 2\frac{F\dot{a}(t)r}{B(\bar{r})^2}A^{\frac{\alpha^2+3\alpha-2}{\alpha}}\right\}dRd\bar{t}$$

$$+ \left\{ \frac{A^2}{B(\bar{r})^2} - \psi^2 \left[A^{2\alpha} - \frac{\dot{a}^2(t)r^2}{B(\bar{r})^2} A^{\frac{2}{\alpha}(\alpha^2 + 2\alpha - 2)} \right] \right.$$

$$\left. +2 \frac{\psi \dot{a}(t)r}{B(\bar{r})^2} A^{\frac{\alpha^2 + 3\alpha - 2}{\alpha}} \right\} dR^2 + R^2 d\Omega_{(2)}^2 . \tag{5.17}$$

By setting

$$\psi = \frac{\dot{a}(t)r}{B^2} \frac{A^{\frac{-\alpha^2 + 3\alpha - 2}{\alpha}}}{D(t, \bar{r})} \tag{5.18}$$

with

$$D(t, \bar{r}) \equiv 1 - \frac{\dot{a}^2(t)r^2}{B^2} A^{\frac{4}{\alpha}(\alpha - 1)} , \tag{5.19}$$

the line element assumes the form [27]

$$ds^2 = -A^{2\alpha}DF^2 d\bar{t}^2 + \left(\frac{H^2}{B^4 D} R^2 A^{2(2-\alpha)} + \frac{A^2}{B^2} \right) dR^2 + R^2 d\Omega_{(2)}^2 , \tag{5.20}$$

where $H \equiv \dot{a}(t)/a(t)$. The apparent horizons, roots of $g^{RR} = 0$, solve

$$\frac{B^4 D}{H^2 R^2 A^{2(2-\alpha)} + A^2 B^2 D} = 0 , \tag{5.21}$$

which reduces to $D = 0$, or

$$B^2 A^{2(\alpha - 1)} = H^2 R^2 . \tag{5.22}$$

Then it must be

$$A^{\alpha - 1} \left[A^2 + \frac{(\alpha - 1)(\alpha + 2)}{\alpha^2} \frac{ma(t)}{R} A^{\frac{(\alpha - 1)(\alpha + 2)}{\alpha}} \right] = \pm HR . \tag{5.23}$$

In an expanding universe, the square bracket is positive and the positive sign is the only one appropriate; Eq. (5.23) becomes

$$HR^2 - \frac{(\alpha - 1)(\alpha + 2)}{\alpha^2} m a(t) A^{\frac{2(\alpha - 1)(\alpha + 1)}{\alpha}} - A^{\alpha + 1}R = 0 . \tag{5.24}$$

If $m > 0$, the Ricci scalar diverges at $R = 0$, identifying a spacetime singularity, at which also $\rho^{(m)}$ is singular.[3]

---

[3]The expression of the Ricci scalar is long and cumbersome and it is not reported here, see Ref. [27].

Let us discuss special limiting cases [27]. When m vanishes (no central object), Eq. (5.24) gives $R = H^{-1}$, the radius of the FLRW Hubble horizon. This value of R is also obtained if $\bar{r} \to +\infty$; then R becomes a comoving radius and the geometry that of spatially flat FLRW space. In this limit Eq. (5.22) shows that $A, B \to 1$ (the limit is not so straightforward in Eq. (5.24) as $R \to \infty$ and $\bar{r} \to \infty$). Based on these features, the horizon at larger radii should be a cosmological one.

Consider now the static limit: when $\beta = 0$, it is $a(t) \equiv a_0$ (Eq. (5.5)). This value of β follows from the choice $\omega_0 = (\gamma - 2)^{-1}$ with $\gamma \neq 2$. For each value of the Brans-Dicke parameter ω_0 there is at most one static Clifton-Mota-Barrow solution corresponding to a specific choice of the equation of state of the cosmic fluid. In order for α to be real, it must be $\omega_0 < -2$ or $\omega_0 > -3/2$, which translates to $\gamma > 3/2$ or $\gamma < 4/3$ when $\beta = 0$.

Equations (5.6) and (5.8) yield

$$\phi(t, R) = \phi_0 \left(\frac{t}{t_0}\right)^2 A^{-\frac{2(\alpha^2-1)}{\alpha}} , \tag{5.25}$$

$$\rho^{(m)} = \rho_0^{(m)} A^{-2\alpha} . \tag{5.26}$$

Even though the metric g_{ab} and the matter density $\rho^{(m)}$ are static, the scalar ϕ depends on time. By writing the line element as

$$ds^2 = -A^{2\alpha} dt^2 + \frac{A^2}{B^2} dR^2 + R^2 d\Omega_{(2)}^2 , \tag{5.27}$$

the apparent horizon equation $g^{RR} = 0$ is $B = 0$, equivalent to the quadratic

$$\bar{r}^2 + \frac{m}{\alpha^2} \left(\alpha^2 - 2\right) \bar{r} + \frac{m^2}{4\alpha^2} = 0 . \tag{5.28}$$

The discriminant $\Delta(\alpha^2) = \frac{m^2}{\alpha^2} \left[\left(\alpha^2 - 2\right)^2 - \alpha^2\right]$ is non-negative if $\alpha \leq 1$ or $\alpha \geq 2$ (using the fact that $\alpha \geq 0$ according to Eq. (5.7)). In the parameter range $1 < \alpha < 2$ the equation $g^{RR} = 0$ has no real roots and there are no apparent horizons. If $\alpha \leq 1$ or $\alpha \geq 2$ there are the real roots

$$\bar{r}_{\pm} = \frac{m}{\alpha^2} \left[-\left(\alpha^2 - 2\right) \pm \sqrt{\left(\alpha^2 - 2\right)^2 - \alpha^2}\right] , \tag{5.29}$$

but they are both negative and no apparent horizon exists: the static spacetime always contains a naked singularity.

Let us discuss now the General Relativity limit $\omega_0 \to \infty$. If $\gamma \neq 0$ and $\gamma \neq 2$, this limit implies $\alpha \to 1$, $\phi \to \phi_0$, and[4]

---

[4]In the case $\gamma = 2$ the scale factor $a(t) \propto \sqrt{t}$, the scalar $\phi \propto t^{-1}$ and the density $\rho^{(m)} \propto t^{-3}$ are independent of ω_0. However, the limit $\omega_0 \to \infty$ still yields $\alpha = 1$ and the various functions of

$$ds^2 = -\left(\frac{1 - \frac{m}{2\bar{r}}}{1 + \frac{m}{2\bar{r}}}\right)^2 dt^2 + a^2(t)\left(1 + \frac{m}{2\bar{r}}\right)^4 \left(d\bar{r}^2 + \bar{r}^2 d\Omega_{(2)}^2\right), \quad (5.30)$$

$$a(t) = a_0 \left(\frac{t}{t_0}\right)^{\frac{2}{3\gamma}}, \quad (5.31)$$

$$\rho^{(m)}(t) = \rho_0^{(m)} \left(\frac{t_0}{t}\right)^2 A^{-2}. \quad (5.32)$$

This line element is recognized as that of a generalized McVittie metric, which becomes

$$ds^2 = -\left(\frac{1 - \frac{M(t)}{2\bar{r}a(t)}}{1 + \frac{M(t)}{2\bar{r}a(t)}}\right)^2 dt^2 + a^2(t)\left(1 + \frac{M(t)}{2\bar{r}a(t)}\right)^4 \left(d\bar{r}^2 + \bar{r}^2 d\Omega_{(2)}^2\right) \quad (5.33)$$

in isotropic coordinates, with $M(t) \geq 0$ an arbitrary function of time and $G_0^1 \neq 0$, corresponding to a radial energy flow. As already seen, the solution with "comoving mass function" $M(t) = M_0 a(t)$ (where M_0 is a constant) is a late-time attractor in the class of generalized McVittie solutions (5.33). This attractor solution is precisely the $\omega_0 \to \infty$ limit of the Clifton-Mota-Barrow solutions (5.2)–(5.8) (which are indeed accreting).[5]

For large values of ω_0, the solution (5.2)–(5.8) asymptotes to the attractor of the generalized McVittie family of solutions.

The $\gamma = 0$ case describes a cosmological constant "fluid"; in this case the exponent β of the scale factor $a(t)$ diverges as $\omega_0 \to \infty$ and this scale factor becomes a power law (the General Relativity limit of the solution is expected to be the Schwarzschild-de Sitter-Kottler spacetime).

If $\gamma = 2$, the $\omega_0 \to \infty$ limit yields $\alpha \to 1$, $\phi \propto t^{-1}$, $a(t) \propto \sqrt{t}$, $\rho^{(m)} \propto t^{-3} A^{-2}$, and the line element is as in Eq. (5.30).

Leaving behind all these special cases, let us inspect now the structure and dynamics of the apparent horizons of the generic Clifton-Mota-Barrow spacetime [27]. It is convenient to introduce the variable $x \equiv \dfrac{m}{2\alpha\bar{r}}$, in terms of which we have

$$A = \frac{1 - x}{1 + x}, \quad (5.34)$$

$\bar{r}$ have the same functional dependence as in the $\gamma \neq 2$ case. The line element is a generalized McVittie one.

[5]The generalized McVittie solutions of General Relativity were introduced 2 years after the Clifton-Mota-Barrow paper, and the attractor solution with $M = M_0 a(t)$ could, in principle, have been found earlier as the limit to General Relativity, but its attractor role follows from different considerations.

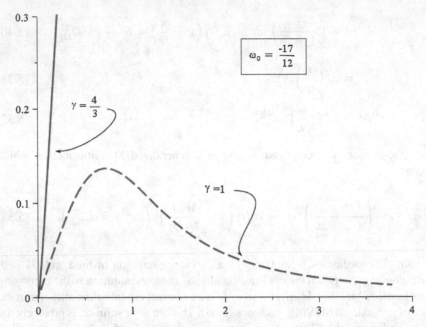

Fig. 5.1 Apparent horizon radii versus time (both in units of $(ma_*)^{1/(1-\beta)}$) for the value $\omega_0 = -17/12$ of the Brans-Dicke parameter. The *dashed curve* corresponds to $\gamma = 1$ (dust) and the *solid curve* to both $\gamma = 4/3$ (radiation) and $\gamma = 2$ (stiff matter). For dust, there is only one apparent horizon which expands to a maximum size and then shrinks. Universes filled with radiation or stiff matter, instead, contain naked singularities

while $H = \beta/t$. The areal radius R of the apparent horizon(s) and the time t can be expressed parametrically as functions of x, obtaining

$$R(x) = a_* t^\beta \frac{m}{2\alpha} \frac{(1+x)^2}{x} \left(\frac{1-x}{1+x}\right)^{\frac{(\alpha-1)(\alpha+2)}{\alpha}} , \tag{5.35}$$

$$t(x) = \left\{ \frac{2\alpha}{m a_* \beta} \frac{x}{(1+x)^{\frac{2}{\alpha}(\alpha+1)}} \left[(1-x)^{2/\alpha} \right.\right.$$

$$\left.\left. +2x \frac{(\alpha-1)(\alpha+2)}{\alpha} (1-x)^{-2(\alpha-1)/\alpha} \right] \right\}^{\frac{1}{\beta-1}} . \tag{5.36}$$

Figures 5.1–5.5 show the areal radii of the apparent horizons versus time for the Brans-Dicke parameter values $\omega_0 = -17/12, -1/3, 1$, and for various large values of ω_0 (of the order of 10^5), respectively, and for various choices of the equation of state parameter γ. In these figures (which follow those of Ref. [27]), R and t are reported in units of

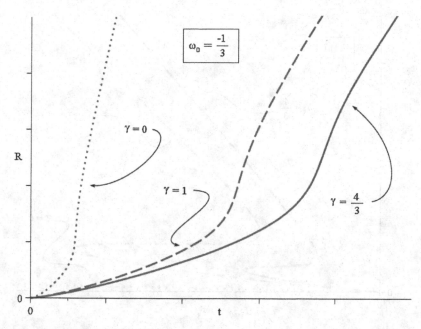

Fig. 5.2 The apparent horizon radii for the Brans-Dicke parameter value $\omega_0 = -1/3$. The *dotted curve* corresponds to $\gamma = 0$ (cosmological constant). In all cases there is a single expanding horizon and the spacetime contains a naked singularity

$$(ma_*)^{\frac{1}{1-\beta}} = \left(a_0 \frac{m}{t_0}\right)^{\frac{1}{1-\beta}} t_0 \tag{5.37}$$

(this normalization absorbs completely the parameters m, a_0, and t_0).

The dotted curves describe $\gamma = 0$ (cosmological constant); the dashed curves correspond to $\gamma = 1$ (dust), while the solid curves correspond to both $\gamma = 4/3$ (radiation) and $\gamma = 2$ (stiff matter). For both values $\gamma = 4/3, 2$, β assumes the value $1/2$ and is independent of ω_0. For the value $\omega_0 = -17/12$ of the Brans-Dicke parameter (Fig. 5.1), a cosmological constant ($\gamma = 0$) gives a contracting universe and this case is not plotted.

For $\omega_0 = -17/12$ and $\omega_0 = -1/3$ there is only one apparent horizon for all of the values of γ explored. In most cases, this horizon is expanding forever and the spacetime contains a naked singularity. For $\omega_0 = -17/12$ and for dust, the apparent horizon expands to a maximum size, stops, and then contracts to zero size asymptotically [27].

When $\omega_0 = 1$ (Figs. 5.3 and 5.4), for dust, radiation, and stiff matter there is initially a single expanding apparent horizon (Fig. 5.3), then two additional apparent horizons appear; the outer horizon expands while the inner horizon eventually approaches the initial one, at which point they merge and disappear, reproducing the "S-curve" phenomenology of the Husain-Martinez-Nuñez solution [27].

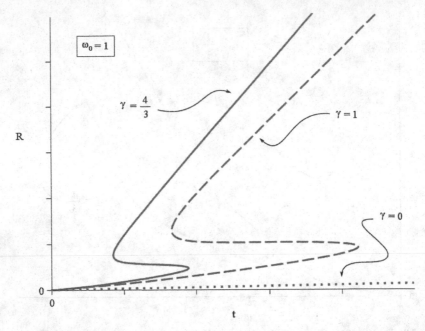

Fig. 5.3 Apparent horizon radii for $\omega_0 = 1$. For all three values of γ, there is a single horizon at early times. As time progresses, two more apparent horizons appear, covering the central singularity. Two of these horizons eventually merge and disappear; then there remains a naked singularity in a FLRW universe, which has its own cosmological horizon. The third curve, flattened along the time axis, is zoomed on in Fig. 5.4

If $\omega_0 = 1$ and $\gamma = 0$ (cosmological constant, Fig. 5.4), one has similar dynamics of the apparent horizons but the new pair of horizons forms inside the original one.

Figure 5.5 corresponds to large values of the parameter ω_0. The apparent horizons exhibit an S-curve behaviour very similar to phenomenology already encountered in solutions of General Relativity [27].

5.2.2 Conformally Transformed Husain-Martinez-Nuñez Spacetime

In addition to the solutions just considered, Clifton, Mota, and Barrow [13] generated another dynamical and spherically symmetric solution of the Brans-Dicke field equations by performing a conformal transformation of the Husain-Martinez-Nuñez metric:

$$g_{ab}^{(\text{HMN})} \longrightarrow \Omega^2 g_{ab}^{(\text{HMN})} = \phi \, g_{ab}^{(\text{HMN})}, \qquad (5.38)$$

$$\phi \longrightarrow \tilde{\phi} = \sqrt{\frac{2\omega + 3}{16\pi}} \ln \phi. \qquad (5.39)$$

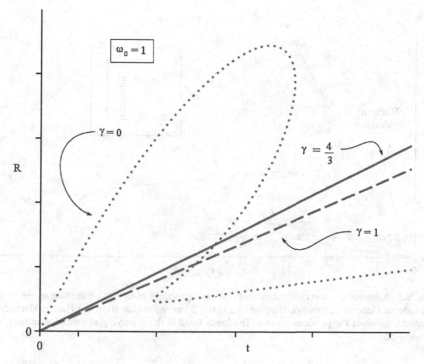

Fig. 5.4 The same as Fig. 5.3, zoomed in

While, in studying scalar-tensor gravity, it is customary to perform the inverse conformal transformation (from the Jordan to the Einstein frame) to end up with an Einstein frame formulation in which the Ricci scalar couples minimally to a scalar field with canonical kinetic energy (but the latter couples non-minimally to all forms of matter which are not trace-free), Ref. [13] transforms a General Relativity metric to a Jordan frame to generate a solution of the Brans-Dicke field equations. The two-parameter geometry thus obtained is described by the line element and scalar field [13]

$$ds^2 = -A^{\alpha\left(1-\frac{1}{\sqrt{3}\beta}\right)}(r)\,dt^2$$

$$+A^{-\alpha\left(1+\frac{1}{\sqrt{3}\beta}\right)}(r)\,t^{\frac{2(\beta-\sqrt{3})}{3\beta-\sqrt{3}}}\left[dr^2 + r^2 A(r)d\Omega_{(2)}^2\right], \tag{5.40}$$

$$\phi(t,r) = A^{\frac{\pm 1}{2\beta}}(r)\,t^{\frac{2}{\sqrt{3}\beta-1}}, \tag{5.41}$$

where

$$A(r) = 1 - \frac{2C}{r}, \quad \beta = \sqrt{2\omega + 3}, \quad \omega > -3/2, \quad \alpha = \pm\sqrt{3}/2. \tag{5.42}$$

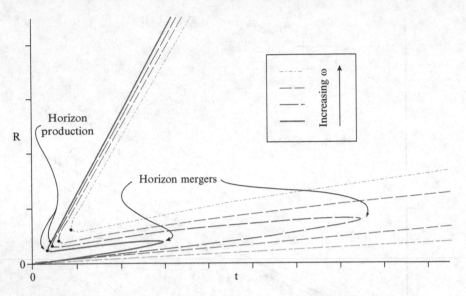

Fig. 5.5 Apparent horizon radii for various (increasing) values of ω_0 (remember that $\omega_0 \to +\infty$ reproduces General Relativity). Here one finds the S-curve familiar from the Husain-Martinez-Nuñez solution of the previous chapter. The lower bend in the S-curve gets pushed at infinity as $\omega \to +\infty$

Spacetime singularities are present at $r = 2C$ and at $t = 0$. The physical coordinate range is $2C < r < +\infty$ and $t > 0$. The result is an inhomogeneous spacetime with a spatially flat FLRW "background" which has scale factor

$$a(t) = t^{\frac{\beta - \sqrt{3}}{3\beta - \sqrt{3}}} \equiv t^{\gamma}. \tag{5.43}$$

Following [26], let us rewrite the line element using the notation and the areal radius

$$ds^2 = -A^{\sigma}(r)\, dt^2 + A^{\Theta}(r)\, a^2(t) dr^2 + R^2(t, r) d\Omega_{(2)}^2, \tag{5.44}$$

$$R(t, r) = A^{\frac{\Theta + 1}{2}}(r)\, a(t)\, r, \tag{5.45}$$

where

$$\sigma = \alpha \left(1 - \frac{1}{\sqrt{3}\,\beta}\right), \tag{5.46}$$

$$\Theta = -\alpha \left(1 + \frac{1}{\sqrt{3}\,\beta}\right). \tag{5.47}$$

Equation (5.45) gives

$$dr = \frac{dR - A^{\frac{\Theta+1}{2}}(r)\dot{a}(t)rdt}{A^{\frac{\Theta-1}{2}}(r)a(t)\frac{C(\Theta+1)}{r} + A^{\frac{\Theta+1}{2}}(r)a(t)} \tag{5.48}$$

which, substituted into the line element, produces

$$ds^2 = \frac{1}{D_1(r)}\left\{-\left[D_1A^\sigma - A^{\frac{\Theta+1}{2}}\dot{a}^2r^2\right]dt^2\right.$$
$$\left.-2A^{\frac{\Theta+1}{2}}\dot{a}r\,dtdR + dR^2\right\} + R^2d\Omega^2_{(2)}, \tag{5.49}$$

with

$$D_1(r) = A(r)\left[1 + \frac{C(\Theta+1)}{rA(r)}\right]. \tag{5.50}$$

Once again, the time-radius cross-term is removed by introducing a new time coordinate T defined by

$$dt = dT - v(t,r)dR \tag{5.51}$$

which, substituted into the line element, yields

$$ds^2 = \frac{1}{D_1}\left[-\left(D_1A^\sigma - A^{\Theta+1}\dot{a}^2r^2\right)dT^2\right.$$
$$+\left(2D_1A^\sigma v - 2vA^{\Theta+1}\dot{a}^2r^2 - 2A^{\frac{\Theta+1}{2}}\dot{a}r\right)dTdR$$
$$\left.+\left(1 + 2A^{\frac{\Theta+1}{2}}\dot{a}r + 2A^{\frac{\Theta+1}{2}}\dot{a}^2r^2 - D_1A^\sigma v^2\right)dR^2\right]$$
$$+R^2d\Omega^2_{(2)}.$$

Let us choose

$$v(t,r) = \frac{A^{\frac{\Theta+1}{2}}\dot{a}r}{D_1A^\sigma - A^{\frac{\Theta+1}{2}}\dot{a}^2r^2} = \frac{H^2R^2}{D_1A^\sigma - H^2R^2}; \tag{5.52}$$

then the line element reduces to

$$ds^2 = \frac{1}{D_1}\left\{-\left(D_1A^\sigma - H^2R^2\right)dT^2\right.$$
$$+\left[1 + 2HR + \left(H^2R^2 - D_1A^\sigma\right)\frac{H^2R^2}{D_1A^\sigma - H^2R^2}\right]\right\}dR^2$$
$$+R^2d\Omega^2_{(2)}. \tag{5.53}$$

There are now two possibilities: either $D_1 A^\sigma - H^2 R^2$ vanishes, or it is different from zero. We consider these two cases separately.

Consider first the situations in which $D_1 A^\sigma \neq H^2 R^2$; then the line element is simply

$$ds^2 = \frac{1}{D_1}\left[-\left(D_1 A^\sigma - H^2 R^2\right) dT^2 + \left(1 + 2HR + H^2 R^2\right)\right] dR^2 + R^2 d\Omega_{(2)}^2. \quad (5.54)$$

The usual recipe $g^{RR} = 0$ locating the apparent horizons is equivalent to $D_1(r) = 0$, which is satisfied by

$$r = (1 - \Theta) C \equiv r_{\text{AH}}. \quad (5.55)$$

Expressed using the areal radius, this equation is

$$R_{\text{AH}}(t) = \left(\frac{\Theta + 1}{\Theta - 1}\right)^2 (1 - \Theta) Ca(t). \quad (5.56)$$

If $\Theta < -1$, this formal root lies in the physical region $r_{\text{AH}} > 2C$, in which the areal radius can be written as

$$R(t, r) = \frac{a(t)r}{(1 - 2C/r)^{\left|\frac{\Theta+1}{2}\right|}}. \quad (5.57)$$

We see that $R \to +\infty$ as $r \to 2C^+$ and, since

$$\frac{\partial R}{\partial r} = a(t) A^{\frac{\Theta-1}{2}}(r)\left[1 - \frac{2C(1 - \Theta)}{2r}\right], \quad (5.58)$$

for $\Theta < -1$ it is $\dfrac{1 - \Theta}{2} > 1$ and $\partial R/\partial r > 0$ if

$$r > r_0 \equiv 2C\left(\frac{1 - \Theta}{2}\right) > 2C, \quad (5.59)$$

while $\partial R/\partial r = 0$ at $r = r_0$, and $\partial R/\partial r < 0$ for $r < r_0$. Because

$$\Theta = \mp\frac{\sqrt{3}}{2}\left(1 + \frac{1}{\sqrt{3}\sqrt{2\omega + 3}}\right) \quad (5.60)$$

for $\alpha = \pm\sqrt{3}/2$, the condition $\Theta < -1$ for the apparent horizon to exist in the physical region requires $\alpha = +\sqrt{3}/2$. The sufficient condition $\Theta < -1$ restricts the Brans-Dicke parameter to

$$\omega < \frac{1}{2} \left[\frac{1}{\left(2 - \sqrt{3}\right)^2} - 3 \right] \equiv \omega_0 , \tag{5.61}$$

hence the apparent horizon exists at $r_0 > 2C$ for the parameter range

$$-\frac{3}{2} < \omega < \omega_0 \tag{5.62}$$

and it has areal radius

$$R_{AH}(t) = \left| \frac{\Theta + 1}{\Theta - 1} \right|^{-\frac{|\Theta + 1|}{2}} |\Theta - 1| a(t) C . \tag{5.63}$$

This apparent horizon is comoving with the cosmic fluid and disappears in the limit $C \to 0$ in which there is no central inhomogeneity.

If $\omega \geq \omega_0$ or if $\alpha = -\sqrt{3}/2$, there is no apparent horizon and the conformal relative of the Husain-Martinez-Nuñez spacetime contains a naked singularity. In particular, when $\alpha = -\sqrt{3}/2$ it is $\Theta = \frac{\sqrt{3}}{2} \left(1 + \frac{1}{\sqrt{3} \beta} \right) > 0$ and the areal radius

$$R(t, r) = \left(1 - \frac{2C}{r} \right)^{\frac{|\Theta + 1|}{2}} a(t) r \tag{5.64}$$

is a monotonically increasing function of r in the range $2C < r < +\infty$.

Let us consider now the case $D_1 A^\sigma = H^2 R^2$; the line element appropriate to this situation is

$$ds^2 = \frac{A^\sigma}{H^2 R^2} \left(-2HR \, dt dR + dR^2 \right) + R^2 d\Omega_{(2)}^2 . \tag{5.65}$$

The inverse metric has components

$$(g^{\mu\nu}) = \begin{pmatrix} -A^\sigma & -\frac{HR}{A^\sigma} & 0 & 0 \\ -\frac{HR}{A^\sigma} & 0 & 0 & 0 \\ 0 & 0 & \frac{1}{R^2} & 0 \\ 0 & 0 & 0 & \frac{1}{R^2 \sin^2 \theta} \end{pmatrix} \tag{5.66}$$

with g^{RR} identically vanishing, hence the condition $D_1 A^\sigma - H^2 R^2 = 0$ imposed from the outset is actually the equation locating the apparent horizons. One can write it as

$$HR = \sqrt{D_1 A^\sigma} = \left[1 + \frac{C(\Theta - 1)}{r} \right] A^{\frac{\sigma - 1}{2}} \tag{5.67}$$

(choosing the positive sign for the square root), where $r = r(T, R)$. In the limit $r \to +\infty$ (and also in the parameter limit $C \to 0$) it is $R \simeq H^{-1}$, the radius of the FLRW apparent horizon. The apparent horizons can be located numerically by using a parametric representation similar to that already seen for the Husain-Martinez-Nuñez spacetime. The result is an "S-curve" completely analogous to that of the Husain-Martinez-Nuñez geometry [26].

5.3 $f(\mathscr{R})$ Cosmological Black Holes

In General Relativity, the current acceleration of the universe [2, 32, 42–46, 54] requires that approximately 73 % of the energy content of the universe be in an exotic form, "dark energy" with pressure $P^{(m)} \sim -\rho^{(m)}$ [1, 33, 34]. As an alternative to this mysterious and ad hoc dark energy, it has been reasoned that perhaps gravity deviates from General Relativity at large scales. A simple model of modified gravity replacing Einstein theory at large scales is $f(\mathscr{R})$ gravity [8, 9, 49, 52, 53, 55], so called from the form of the action

$$S = \int d^4x \sqrt{-g} \left[f(\mathscr{R}) + \mathscr{L}^{(m)} \right] \tag{5.68}$$

which reduces to the Einstein-Hilbert action for a linear function f of the Ricci scalar $\mathscr{R}$. This modified gravity can in principle explain the cosmic acceleration and be theoretically consistent [7, 15, 50]. Since $f(\mathscr{R})$ theories interesting for cosmology contain a time-varying effective cosmological "constant", black holes or local objects are dynamical and asymptotically FLRW, not asymptotically flat. Very few analytic solutions of this kind are known.

A rare spherically symmetric dynamical solution in vacuum $f(\mathscr{R}) = \mathscr{R}^{1+\delta}$ gravity was found by Clifton [10] and studied in [21]. The parameter δ of $\mathscr{R}^{1+\delta}$ gravity is severely constrained by Solar System experiments to be in the range $\delta = (-1.1 \pm 1.2) \cdot 10^{-5}$ [3, 11, 12, 59]. The *Clifton line element* is

$$ds^2 = -A_2(\bar{r})dt^2 + a^2(t)B_2(\bar{r}) \left(d\bar{r}^2 + \bar{r}^2 d\Omega_{(2)}^2 \right) , \tag{5.69}$$

where

$$A_2(\bar{r}) = \left(\frac{1 - C_2/\bar{r}}{1 + C_2/\bar{r}} \right)^{2/q} , \tag{5.70}$$

$$B_2(\bar{r}) = \left(1 + \frac{C_2}{\bar{r}} \right)^4 A_2(\bar{r})^{q+2\delta-1} , \tag{5.71}$$

$$a(t) = t^{\frac{\delta(1+2\delta)}{1-\delta}} , \tag{5.72}$$

$$q^2 = 1 - 2\delta + 4\delta^2 , \tag{5.73}$$

in isotropic coordinates. Equation (5.69) gives back the FLRW line element when the mass parameter C_2 vanishes. This modified gravity reduces to General Relativity when $\delta \to 0$, in which the metric (5.69) reproduces the Schwarzschild solution in isotropic coordinates. We assume that C_2 is positive and we assume that $\delta > 0$ for stability (in fact, local stability of the theory requires $f''(\mathcal{R}) \geq 0$ [18, 19]).

Clifton's solution (5.69)–(5.73) is conformal to the Fonarev solution of General Relativity already seen and, since the latter is conformally static [35], also the Clifton solution is.

A first transformation to the radial coordinate

$$r \equiv \bar{r}\left(1 + \frac{C_2}{\bar{r}}\right)^2 , \tag{5.74}$$

in terms of which $d\bar{r} = \left(1 - \frac{C_2^2}{\bar{r}^2}\right)^{-1} dr$, followed by another transformation to the areal radius

$$R \equiv \frac{a(t)\sqrt{B_2(\bar{r})}\, r}{\left(1 + \frac{C_2}{\bar{r}}\right)^2} = a(t)\, r A_2(\bar{r})^{\frac{q+2\delta-1}{2}} . \tag{5.75}$$

brings the line element (5.69) to the form [21]

$$ds^2 = -A_2 dt^2 + a^2 A_2^{2\delta-1} dr^2 + R^2 d\Omega_{(2)}^2 . \tag{5.76}$$

Now we have

$$dr = \frac{dR - A_2^{\frac{q+2\delta-1}{2}}\,\dot{a}\,r\,dt}{a\left[A_2^{\frac{q+2\delta-1}{2}} + \frac{2(q+2\delta-1)}{q}\frac{C_2}{r}A_2^{\frac{2\delta-1-q}{2}}\right]} \equiv \frac{dR - A_2^{\frac{q+2\delta-1}{2}}\,\dot{a}\,r\,dt}{aA_2^{\frac{q+2\delta-1}{2}}\,C(\bar{r})} . \tag{5.77}$$

(where an overdot denotes differentiation with respect to the comoving time t of the FLRW "background") and

$$C(\bar{r}) = 1 + \frac{2(q+2\delta-1)}{q}\frac{C_2}{r}A_2^{-q} = 1 + \frac{2(q+2\delta-1)}{q}\frac{C_2 a}{R}A_2^{\frac{2\delta-1-q}{2}} . \tag{5.78}$$

The metric is recast as

$$ds^2 = -A_2\left[1 - \frac{A_2^{2(\delta-1)}}{C^2}\dot{a}^2 r^2\right]dt^2 - \frac{2A_2^{\frac{-q+2\delta-1}{2}}}{C^2}\dot{a}\,r\,dtdR$$

$$+ \frac{dR^2}{A_2^q C^2} + R^2 d\Omega_{(2)}^2 . \tag{5.79}$$

The cross-term in $dtdR$ is eliminated by replacing t with another time coordinate $\bar{t}$ which satisfies

$$d\bar{t} = \frac{1}{F(t,R)}\left[dt + \beta(t,R)dR\right],\qquad(5.80)$$

where $F(t,R)$ is an integrating factor chosen to make $d\bar{t}$ exact. In terms of this time coordinate, the line element assumes the form [21]

$$ds^2 = -A_2\left[1 - \frac{A_2^{2(\delta-1)}}{C^2}\dot{a}^2r^2\right]F^2d\bar{t}^2$$

$$+2F\left\{A_2\beta\left[1 - \frac{A_2^{2(\delta-1)}}{C^2}\dot{a}^2r^2\right] - \frac{A_2^{\frac{-q+2\delta-1}{2}}}{C^2}\dot{a}r\right\}d\bar{t}dR$$

$$+\left\{-A_2\left[1 - \frac{A_2^{2(\delta-1)}}{C^2}\dot{a}^2r^2\right]\beta^2 + \frac{2A_2^{\frac{-q+2\delta-1}{2}}}{C^2}\dot{a}r\beta + \frac{1}{A_2^qC^2}\right\}dR^2$$

$$+R^2d\Omega_{(2)}^2.\qquad(5.81)$$

The choice of β

$$\beta = \frac{A_2^{\frac{-q+2\delta-3}{2}}}{C^2}\frac{\dot{a}r}{1 - \frac{A_2^{2(\delta-1)}}{C^2}\dot{a}^2r^2}\qquad(5.82)$$

removes the unwanted cross-term and leaves the geometry in the final form

$$ds^2 = -A_2DF^2d\bar{t}^2 + \frac{1}{A_2^qC^2}\left[1 + \frac{A_2^{-q-1}H^2R^2}{C^2D}\right]dR^2 + R^2d\Omega_{(2)}^2,\qquad(5.83)$$

where $H \equiv \dot{a}/a$ and

$$D \equiv 1 - \frac{A_2^{2(\delta-1)}}{C^2}\dot{a}^2r^2 = 1 - \frac{A_2^{-q-1}}{C^2}H^2R^2.\qquad(5.84)$$

Further manipulation yields [21]

$$ds^2 = -A_2DF^2d\bar{t}^2 + \frac{dR^2}{A_2^qC^2D} + R^2d\Omega_{(2)}^2.\qquad(5.85)$$

The equation $g^{RR} = 0$ locating the apparent horizons holds if $A_2^qC^2D = 0$, giving

$$A_2^q \left(C^2 - H^2 R^2 A_2^{-q-1} \right) = 0 , \tag{5.86}$$

so the apparent horizons are located by $A_2 = 0$ or $H^2 R^2 = C^2 A_2^{q+1}$. The expression A_2 goes to zero for $\bar{r} = C_2$, which describes the Schwarzschild event horizon in the limit to General Relativity $\delta \to 0$. Now $A_2 = 0$ identifies a singularity because the Ricci scalar

$$\mathscr{R} = \frac{6 \left(\dot{H} + 2H^2 \right)}{A_2(\bar{r})} \tag{5.87}$$

diverges as $\bar{r} \to C_2$.

The second possibility to satisfy $g^{RR} = 0$ gives $H^2 R^2 = C^2 A_2^{q+1}$, which translates to

$$HR = \pm \left[1 + \frac{2(q + 2\delta - 1)}{q} \frac{C_2 a}{R} A_2^{\frac{2\delta-1-q}{2}} \right] A_2^{\frac{q+1}{2}} , \tag{5.88}$$

discarding the negative sign in an expanding universe. In the limit $\delta \to 0$ Eq. (5.88) reduces to

$$HR = \left[1 + \frac{2\delta C_2 a}{R} A_2^{-\left(1 - \frac{3\delta}{2}\right)} \right] A_2^{1-\delta} . \tag{5.89}$$

It is instructive to discuss two limits. In the first limit, $C_2 \to 0$ and the central object disappears, leaving behind FLRW space; then the coordinate $\bar{r} = r$ approaches the comoving radius of FLRW space and R approaches the proper radius of this space. Equation (5.88) then degenerates into $R_c = 1/H$ (the FLRW cosmological horizon). In the General Relativity limit $\delta \to 0$, Eq. (5.88) reduces to $\bar{r} = C_2$ with $H \equiv 0$.

Using $x \equiv C_2/\bar{r}$, Eqs. (5.72) and (5.75) allow one to write the left hand side of (5.88) as

$$HR = \frac{\delta (1 + 2\delta)}{1 - \delta} t^{\frac{2\delta^2 + 2\delta - 1}{1 - \delta}} \frac{C_2}{x} \frac{(1 - x)^{\frac{q + 2\delta - 1}{q}}}{(1 + x)^{\frac{-q + 2\delta - 1}{q}}} , \tag{5.90}$$

while the right hand side of Eq. (5.88) is

$$\frac{(1 - x)^{\frac{q+1}{q}}}{(1 + x)^{\frac{q+1}{q}}} \left[1 + \frac{2 (q + 2\delta - 1)}{q} \frac{x}{(1 - x)^2} \right] \tag{5.91}$$

and Eq. (5.88) is written as[6]

---

[6]Here $\dfrac{1 - 2\delta - 2\delta^2}{1 - \delta} > 0$ for $0 < \delta < \dfrac{\sqrt{3} - 1}{2} \simeq 0.366$.

$$\frac{1}{t^{\frac{1-2\delta-2\delta^2}{1-\delta}}} = \frac{(1-\delta)}{\delta\,(1+2\delta)\,C_2}\,\frac{x\,(1+x)^{\frac{-2q+2\delta-2}{q}}}{(1-x)^{\frac{2(\delta-1)}{q}}}$$

$$\cdot\left[1+\frac{2\,(q+2\delta-1)}{q}\,\frac{x}{(1-x)^2}\right]. \tag{5.92}$$

The left hand side of Eq. (5.92) vanishes at late times, and $x \simeq 0$; then there is a single root of $g^{RR} = 0$, identified with the radius of a cosmological apparent horizon ($\bar{r} \to \infty$ as $x = C_2/\bar{r} \to 0$ and the limit $x \to 0$ can also be obtained as the parameter $C_2 \to 0$, in which case $HR \to 1$ and $\bar{r} \simeq R \simeq H^{-1} = \dfrac{1-\delta}{\delta\,(1+2\delta)}\,t$). At late times there is only one cosmological apparent horizon and the Clifton spacetime contains a naked singularity at $R = 0$.

A parametric representation of the apparent horizon radius R and time t is

$$R(x) = t(x)^{\frac{\delta(1+2\delta)}{1-\delta}}\,\frac{C_2}{x}\,(1-x)^{\frac{q+2\delta-1}{q}}\,(1+x)^{\frac{q-2\delta+1}{q}}\,, \tag{5.93}$$

$$t(x) = \left\{\frac{(1-\delta)}{\delta\,(1+2\delta)\,C_2}\,\frac{x\,(1+x)^{\frac{2(-q+\delta-1)}{q}}}{(1-x)^{\frac{2(\delta-1)}{q}}}\left[1+\frac{2\,(q+2\delta-1)\,x}{q(1-x)^2}\right]\right\}^{\frac{1-\delta}{2\delta^2+2\delta-1}} \tag{5.94}$$

The qualitative behaviour of the apparent horizon radii is described by an "S-curve" similar to that of the Husain-Martinez-Nuñez spacetime [21].

5.4 Conclusions

To end these lectures, we are now aware of several explicit examples of time-varying apparent horizons in General Relativity and in alternative theories of gravity. These examples will be useful to study Hawking radiation and black hole thermodynamics in fully dynamical situations (and in part, they are already beginning to be used for this purpose [16, 20, 24, 25, 37–39, 47]). Many aspects of fundamental gravitational physics are touched upon in the study of apparent horizons: perhaps the most obvious is the long-standing issue of cosmological expansion versus local dynamics, which prompted McVittie to produce his solution of the Einstein equations in 1933. To turn things around, one could as well say that another aspect is present, namely the effect of a central inhomogeneity on the cosmological expansion of a "background" universe (although it is the first effect that is usually emphasized).

Although the McVittie geometry has been largely overlooked and does not make it to the relativity textbooks, it has seen a resurgence of interest in recent years and it has been shown to be a solution also of very interesting modern theories, such as Hořava-Lifshitz gravity, Horndeski theory, and shape dynamics.

Analytic solutions of the field equations describing cosmological black holes allow us to study the effect of "normal" versus phantom backgrounds on black holes, especially the accretion of dark energy (and of its extreme form, phantom energy) by black holes, which has been the subject of a significant amount of literature. There is, however, an intrinsic limitation in the test fluid approximation used in the literature to address this problem, and only exact solutions can provide some answer to the various questions which go unanswered.

The scope enlarges when one attempts to go beyond General Relativity. Alternative theories of gravity may be required already to explain the current acceleration of the universe without dark energy. A huge amount of literature has been devoted to $f(\mathscr{R})$ gravity (which is a special class of scalar-tensor theories with a complicated scalar field potential) for this purpose.[7] In $f(\mathscr{R})$ and scalar-tensor gravity designed for cosmology, the theory contains an effective time-dependent cosmological "constant" and black holes in these theories are asymptotically FLRW, not asymptotically flat, and they are dynamical (except for the special case of a de Sitter "background"). Only a few analytic solutions of these theories describing cosmological black holes are known, and they exhibit various phenomenologies of apparent horizons.

In the context of scalar-tensor gravity, cosmological black holes provide toy models to study the spatial variation of fundamental constants (for example, the variation of the gravitational coupling G_{eff}). Containing only a scalar extra degree of freedom, scalar-tensor gravity is a minimal modification of Einstein's theory. As such, it is justly regarded as the prototypical alternative to Einstein's theory [57] and it was quite interesting to discuss cosmological black hole solutions of the relevant field equations. The fact that such a variety of behaviours (cosmological black holes, naked singularities, appearing/bifurcating and merging/disappearing pairs of apparent horizons) is contained in this relatively simple theory of gravity induces the suspicion that more complicated theories of gravity will disclose higher degrees of richness and complication of non-stationary horizons, which have not yet been unveiled.

As a general consideration, while it is necessary to find new solutions of the field equations containing apparent and trapping time-evolving horizons, and it is good to extend the catalogue, it is more important to understand the known solutions. At least, this is the lesson that one draws from the history of the McVittie solution. For certain analytic dynamical solutions of well known field equations, it is not even known if they represent black holes or naked singularities, and probably new apparent horizon phenomenology is waiting to be discovered and fully understood.

The types of questions that should be investigated, and that we tried to address, in the general theory and in the various specific spacetimes discussed, include:

- Do these geometries contain spacetime singularities? In relation with Cosmic Censorship, are singularities naked, or are they dressed by some kind of horizon? Are singularities spacelike, timelike, null, and do they change their causal

---

[7]At the moment of writing, a crude estimate is 2000 theoretical and observational journal articles devoted to $f(\mathscr{R})$ gravity since 2003.

character? The implication of timelike naked singularities is that the spacetime cannot be derived by regular Cauchy data and the world becomes unpredictable.

- Are there apparent/trapping horizons and, if so, where are they? Are these apparent horizons timelike, spacelike, or null?
- Do apparent horizons in a FLRW "background" expand? How? Are they comoving? For spherically symmetric spacetimes in General Relativity, does the Misner-Sharp-Hernandez mass enclosed by the horizon increase? If possible, the contributions from the mass-energy of the local object, the accretion of cosmic fluid, the expansion/contraction of the apparent horizon, and the time evolution of the energy density of the "background" should be identified and separated.
- What are the dynamics and phenomenology of the apparent horizons? A single apparent horizon may originate from a spacetime singularity (as in the Husain-Martinez-Nuñez solution of General Relativity); apparent horizons are known to appear and disappear in pairs (as in the McVittie and in Lemaître-Tolman-Bondi models); and they appear to jump. This variation with time of the number of apparent horizons is sometimes interpreted as a single trapping horizon tube which goes back and forth in time, generating what looks like the appearance, disappearance, or bifurcation of apparent horizons when the tube is sliced with a hypersurface of constant time. The behaviour of apparent horizons can be much richer than the simple McVittie or "S-curve" phenomenologies, as shown by the Clifton-Mota-Barrow solutions of Brans-Dicke theory which, when zoomed at closely, reveal unintuitive behaviour of their apparent horizons and a richer phenomenology as a wider parameter space is spanned.
- What are the conformal diagrams for these spacetimes? Their construction, and the detailed analysis of the causal structure of spacetime is not, in general, an easy task and has taken almost 80 years to be analyzed in detail for the McVittie space. Since there is freedom in fixing the FLRW "background" for cosmological black holes, there is also a substantial range of possibilities for the causal structure.
- What are the matter sources for these solutions of the relevant field equations? In these lectures we have encountered a single perfect fluid, a mixture of two non-interacting perfect fluids, canonical and phantom scalar fields (free and with exponential potentials), cuscuton fields in Hořava-Lifshitz theory, scalars from Horndeski theory, a mixture of a perfect fluid and a Brans-Dicke field, and pure vacuum geometry in $f(\mathscr{R})$ gravity.

Two questions stand out and remain largely unanswered:

- Are apparent and trapping horizons the right constructs to use in the study of dynamical black holes, as theoretical laboratories in general and for astrophysical purposes? de facto, they are the objects used in numerical relativity to predict accurate waveforms for gravitational wave detection experiments. However, it is possible that better concepts of "horizon" will be discovered in the future.
- It is not established beyond doubt that a thermodynamics of apparent horizons is meaningful. It is not clear why there should be a thermodynamics of equilibrium at all in situations in which these horizons are changing fast (in comparison with some given time scale, for example of a "background"). However, in slowly

evolving situations, it is certainly plausible that a thermodynamics of quasi-equilibrium can be formulated in an adiabatic approximation. More work is needed to establish once and for all the correct concept of horizon temperature for apparent horizons. While, for stationary black holes, various independent methods provide the same result (the Hawking temperature), we do not have the luxury of comparing results computed with different methods for apparent horizons.

Due to the many questions still open, these lectures end without giving a complete view of the field of apparent and trapping horizons because there isn't one yet. This subject is an active area of research and it is hoped that, although the research is sometimes difficult and there are not many clues to follow at the moment of writing, these notes will soon be outdated by new advances in this area.

References

1. Amendola, L., Tsujikawa, S.: Dark Energy, Theory and Observations. Cambridge University Press, Cambridge (2010)
2. Barris, B., et al.: Twenty-three high-redshift supernovae from the Institute for Astronomy Deep Survey: doubling the supernova sample at $z > 0.7$. Astrophys. J. **602**, 571 (2004)
3. Barrow, J.D., Clifton, T.: Exact cosmological solutions of scale-invariant gravity theories. Class. Quantum Grav. **23**, L1 (2006)
4. Bergmann, P.G.: Comments on the scalar tensor theory. Int. J. Theor. Phys. **1**, 25 (1968)
5. Brans, C., Dicke, R.H.: Mach's principle and a relativistic theory of gravitation. Phys. Rev. **124**, 925 (1961)
6. Callan, C.G., Friedan, D., Martinez, E.J., Perry, M.J.: Strings in background fields. Nucl. Phys. B **262**, 593 (1985)
7. Capozziello, S., Faraoni, V.: Beyond Einstein Gravity, a Survey of Gravitational Theories for Cosmology and Astrophysics. Springer, New York (2010)
8. Capozziello, S., Carloni, S., Troisi, A.: Quintessence without scalar fields. Recent Res. Dev. Astron. Astrophys. **1**, 625 (2003)
9. Carroll, S.M., Duvvuri, V., Trodden, M., Turner, M.S.: Is cosmic speed-up due to new gravitational physics? Phys. Rev. D **70**, 043528 (2004)
10. Clifton, T.: Spherically symmetric solutions to fourth-order theories of gravity. Class. Quantum Grav. **23**, 7445 (2006)
11. Clifton, T., Barrow, J.D.: The power of general relativity. Phys. Rev. D **72**, 103005 (2005)
12. Clifton, T., Barrow, J.D.: Class. Quantum Grav. **23**, 2951 (2005)
13. Clifton, T., Mota, D.F., Barrow, J.D.: Inhomogeneous gravity. Mon. Not. R. Astr. Soc. **358**, 601 (2005)
14. Cognola, G., Gorbunova, O., Sebastiani, L., Zerbini, S.: On the energy issue for a class of modified higher order gravity black hole solutions. Phys. Rev. D **84**, 023515 (2011)
15. De Felice, A., Tsujikawa, S.: $f(R)$ theories. Living Rev. Relat. **13**, 3 (2010)
16. Deruelle, N., Sasaki, M.: In: Odintsov, S.D., Sáez-Gómez, D., Xambó-Descamps, S. (eds.) Proceedings, Cosmology, the Quantum Vacuum, and Zeta Functions, Barcelona, 8–10 Mar 2010. Springer Proceedings in Physics, vol. 137, p. 247. Springer, Berlin/New York (2011)
17. Faraoni, V.: Cosmology in Scalar-Tensor Gravity. Kluwer Academic, Dordrecht (2004)
18. Faraoni, V.: Matter instability in modified gravity. Phys. Rev. D **74**, 104017 (2006)
19. Faraoni, V.: Phys. Rev. D **75**, 067302 (2007)

20. Faraoni, V.: Hawking temperature of expanding cosmological black holes. Phys. Rev. D **76**, 104042 (2007)
21. Faraoni, V.: Clifton's spherical solution in f(R) vacuum harbours a naked singularity. Class. Quantum Grav. **26**, 195013 (2009)
22. Faraoni, V.: Black hole entropy in scalar-tensor and $f(R)$ gravity: an overview. Entropy **12**, 1246 (2010)
23. Faraoni, V.: Jebsen-Birkhoff theorem in alternative gravity. Phys. Rev. D **81**, 044002 (2010)
24. Faraoni, V., Nielsen, A.B.: Quasi-local horizons, horizon-entropy, and conformal field redefinitions. Class. Quantum Grav. **28**, 175008 (2011)
25. Faraoni, V., Vitagliano, V.: Horizon thermodynamics and spacetime mappings. Phys. Rev. D **89**, 064015 (2014)
26. Faraoni, V., Zambrano Moreno, A.F.: Interpreting the conformal cousin of the Husain-Martinez-Nuñez solution. Phys. Rev. D **86**, 084044 (2012)
27. Faraoni, V., Vitagliano, V., Sotiriou, T.P., Liberati, S.: Dynamical apparent horizons in inhomogeneous Brans-Dicke universes. Phys. Rev. D **86**, 064040 (2012)
28. Fradkin, E.S., Tseytlin, A.A.: Quantum string theory effective action. Nucl. Phys. B **261**, 1 (1985)
29. Fujii, Y., Maeda, K.: The Scalar-Tensor Theory of Gravitation. Cambridge University Press, Cambridge (2003)
30. Green, M.B., Schwarz, G.H., Witten, E.: Superstring Theory. Cambridge University Press, Cambridge (1987)
31. Hayward, S.A.: Quasilocal gravitational energy. Phys. Rev. D **49**, 831 (1994)
32. Knop, R., et al.: New constraints on Ω_M, Ω_Λ, and w from an independent set of 11 high-redshift supernovae observed with the Hubble Space Telescope. Astrophys. J. **598**, 102 (2003)
33. Komatsu, E., et al.: Seven-year Wilkinson Microwave Anisotropy Probe (WMAP*) observations: cosmological interpretation. Astrophys. J. (Suppl.) **192**, 18 (2011)
34. Linder, E.V.: Resource Letter DEAU-1: Dark energy and the accelerating universe. Am. J. Phys. **76**, 197 (2008)
35. Maeda, H.: Global structure and physical interpretation of the Fonarev solution for a scalar field with exponential potential. Preprint arXiv:0704.2731
36. Maeda, K., Ohta, N., Uzawa, K.: Dynamics of intersecting brane systems-Classification and their applications. J. High Energy Phys. **0906**, 051 (2009)
37. Majhi, B.R.: Thermodynamics of Sultana-Dyer black hole. J. Cosmol. Astropart. Phys. **1405**, 014 (2014)
38. Majhi, B.R.: Conformal transformation, near horizon symmetry, Virasoro algebra, and entropy. Phys. Rev. D **90**, 044020 (2014)
39. Nielsen, A.B., Firouzjaee, J.T.: Conformally rescaled spacetimes and Hawking radiation. Gen. Rel. Gravit. **45**, 1815 (2013)
40. Nordtvedt, K.: PostNewtonian metric for a general class of scalar tensor gravitational theories and observational consequences. Astrophys. J. **161**, 1059 (1970)
41. Nozawa, M., Maeda, H.: Dynamical black holes with symmetry in Einstein-Gauss-Bonnet gravity. Class. Quantum Grav. **25**, 055009 (2008)
42. Perlmutter, S., et al.: Discovery of a supernova explosion at half the age of the Universe. Nature **391**, 51 (1998)
43. Riess, A.G., et al.: Observational evidence from supernovae for an accelerating universe and a cosmological constant. Astron. J. **116**, 1009 (1998)
44. Riess, A.G., et al.: An indication of evolution of type Ia supernovae from their risetimes. Astron. J. **118**, 2668 (1999)
45. Riess, A.G., et al.: The farthest known supernova: support for an accelerating universe and a glimpse of the epoch of deceleration. Astrophys. J. **560**, 49 (2001)
46. Riess, A.G., et al.: Type Ia supernova discoveries at $z > 1$ from the Hubble Space Telescope: evidence for past deceleration and constraints on dark energy evolution. Astron. J. **607**, 665 (2004)

47. Saida, H., Harada, T., Maeda, H.: Black hole evaporation in an expanding universe. Class. Quantum Grav. **24**, 4711 (2007)
48. Sakai, N., Barrow, J.D.: Cosmological evolution of black holes in Brans-Dicke gravity. Class. Quantum Grav. **18**, 4717 (2001)
49. Sotiriou, T.P.: $f(R)$ gravity and scalar-tensor theory. Class. Quantum Grav. **23**, 5117 (2006)
50. Sotiriou, T.P., Faraoni, V.: $f(R)$ theories of gravity. Rev. Mod. Phys. **82**, 451 (2010)
51. Sotiriou, T.P., Faraoni, V.: Black holes in scalar-tensor gravity. Phys. Rev. Lett. **108**, 081103 (2012)
52. Sotiriou, T.P., Liberati, S.: Metric-affine $f(R)$ theories of gravity. Ann. Phys. (N.Y.) **322**, 935 (2007)
53. Sotiriou, T.P., Liberati, S.: The metric-affine formalism of $f(R)$ gravity. J. Phys. Conf. Ser. **68**, 012022 (2007)
54. Tonry, J.L., et al.: Cosmological results from high-z supernovae. Astrophys. J. **594**, 1 (2003)
55. Vollick, D.N.: Phys. Rev. D **68**, 063510 (2003)
56. Wagoner, R.V.: Scalar-tensor theory and gravitational waves. Phys. Rev. D **1**, 3209 (1970)
57. Will, C.M.: Theory and Experiment in Gravitational Physics, 2nd edn. Cambridge University Press, Cambridge (1993)
58. Wiltshire, D.L.: Spherically symmetric solutions of Einstein-Maxwell theory with a Gauss-Bonnet term. Phys. Lett. B **169**, 36 (1986)
59. Zakharov, A.F., Nucita, A.A., De Paolis, F., Ingrosso, G.: Solar system constraints on R^n gravity. Phys. Rev. D **74**, 107101 (2006)

Appendix

> *I have been impressed with the urgency of doing. Knowing is not enough; we must apply. Being willing is not enough; we must do.*
>
> —Leonardo da Vinci

A.1 Painlevé-Gullstrand Coordinates for General Spherically Symmetric Metrics

Beginning from the metric given by Eq. (2.94) and following the notations of Ref. [1], we search for a new time coordinate τ (Painlevé-Gullstrand time). The transformation $t \to \tau(t, R)$ yields

$$d\tau = \frac{\partial \tau}{\partial t} dt + \frac{\partial \tau}{\partial R} dR$$

and

$$dt = \frac{1}{\partial \tau / \partial t} \left(d\tau - \frac{\partial \tau}{\partial R} dR \right),$$

which transforms the line element (2.94) into

$$ds^2 = -\frac{e^{-2\phi} (1 - 2M/R)}{(\partial \tau / \partial t)^2} d\tau^2 + 2e^{-2\phi} \left(1 - \frac{2M}{R} \right) \frac{\partial \tau / \partial R}{(\partial \tau / \partial t)^2} d\tau dR$$

$$+ \left[-e^{-2\phi} \left(1 - \frac{2M}{R} \right) \left(\frac{\partial \tau / \partial R}{\partial \tau / \partial t} \right)^2 + \frac{1}{1 - 2M/R} \right] dR^2 + R^2 d\Omega_{(2)}^2 .$$

$$(A.1)$$

We now impose that the new time coordinate τ is such that $g_{11} = 1$, which implies that

© Springer International Publishing Switzerland 2015
V. Faraoni, *Cosmological and Black Hole Apparent Horizons*,
Lecture Notes in Physics 907, DOI 10.1007/978-3-319-19240-6

$$\frac{\partial \tau}{\partial R} = \pm \frac{e^{\phi}}{1 - 2M/R} \sqrt{\frac{2M}{R}} \frac{\partial \tau}{\partial t}. \tag{A.2}$$

Then, the metric component g_{01} in the new coordinates is

$$g_{01} = \pm \frac{e^{-\phi}}{\partial \tau/\partial t} \sqrt{\frac{2M}{R}} \tag{A.3}$$

and the line element assumes the form (2.95).

A.2   Kodama Vector in FLRW Space

Here we compute the components of the Kodama vector in FLRW space in pseudo-Painlevé-Gullstrand and in comoving coordinates.

A.2.1   Pseudo-Painlevé-Gullstrand Coordinates

In these coordinates the 2-metric h_{ab} of Eq. (2.71) and its inverse are given by

$$(h_{ab}) = \begin{pmatrix} \frac{-\left(1 - H^2 R^2 - kR^2/a^2\right)}{1 - kR^2/a^2} & \frac{-HR}{1 - kR^2/a^2} \\ \frac{-HR}{1 - kR^2/a^2} & \frac{1}{1 - kR^2/a^2} \end{pmatrix}, \tag{A.4}$$

$$(h^{ab}) = \begin{pmatrix} -1 & -HR \\ -HR & \left(1 - H^2 R^2 - kR^2/a^2\right) \end{pmatrix} \tag{A.5}$$

by decomposing the metric (3.25). The volume form on the normal 2-space is

$$\epsilon_{ab} = \sqrt{|h|} \, (dt)_a \wedge (dR)_b = \frac{1}{\sqrt{1 - kR^2/a^2}} \left(\delta_{a0}\delta_{b1} - \delta_{a1}\delta_{b0}\right), \tag{A.6}$$

while

$$\begin{aligned} \epsilon^{ab} &= g^{ac} g^{bd} \frac{\left(\delta_{c0}\delta_{d1} - \delta_{c1}\delta_{d0}\right)}{\sqrt{1 - kR^2/a^2}} \\ &= \frac{\left(h^{a0}h^{b1} - h^{a1}h^{b0}\right)}{\sqrt{1 - kR^2/a^2}}. \end{aligned}$$

The Kodama vector is

$$K^a \equiv \epsilon^{ab} \nabla_b R = \frac{\left(h^{a0} h^{b1} - h^{a1} h^{b0}\right)}{\sqrt{1 - kR^2/a^2}} \delta_{b1}$$

$$= \frac{\left(h^{a0} h^{11} - h^{a1} h^{10}\right)}{\sqrt{1 - kR^2/a^2}}$$

and, therefore,

$$K^0 = \frac{-1}{\sqrt{1 - kR^2/a^2}} \left(1 - H^2 R^2 - kR^2/a^2 + H^2 R^2\right)$$

$$= \frac{-\left(1 - kR^2/a^2\right)}{\sqrt{1 - kR^2/a^2}} = -\sqrt{1 - kR^2/a^2},$$

$$K^1 = \frac{\left(h^{10} h^{11} - h^{11} h^{10}\right)}{\sqrt{1 - kR^2/a^2}} = 0.$$

To conclude, we have

$$K^\mu = \left(-\sqrt{1 - kR^2/a^2}, 0, 0, 0\right) \quad \text{(pseudo-Painlevé-Gullstrand coordinates)}.$$

$$(A.7)$$

A.2.2   Comoving Coordinates

In comoving coordinates the FLRW line element is

$$ds^2 = -dt^2 + \frac{a^2(t)}{1 - kr^2} dr^2 + R^2 d\Omega_{(2)}^2 = h_{ab} dx^a dx^b + R^2 d\Omega_{(2)}^2, \qquad (A.8)$$

where $R = a(t)r$ is the areal radius. The volume form on the 2-space (t, r) has components

$$\epsilon_{\alpha\beta} = \sqrt{|h|} \, (dt)_\alpha \wedge (dr)_\beta = \frac{a}{\sqrt{1 - kr^2}} \left(\delta_{\alpha 0} \delta_{\beta 1} - \delta_{\alpha 1} \delta_{\beta 0}\right)$$

while

$$\epsilon^{\alpha\beta} = g^{\alpha\gamma} g^{\beta\delta} \epsilon_{\gamma\delta} = \frac{a}{\sqrt{1 - kr^2}} g^{\alpha\gamma} g^{\beta\delta} \left(\delta_{\gamma 0} \delta_{\delta 1} - \delta_{\gamma 1} \delta_{\delta 0}\right)$$

$$= \frac{a}{\sqrt{1 - kr^2}} \left(g^{\alpha 0} g^{\beta 1} - g^{\alpha 1} g^{\beta 0}\right).$$

The components of the Kodama vector in comoving coordinates are

$$K^\alpha \equiv \epsilon^{\alpha\beta} \nabla_\beta R = \epsilon^{\alpha\beta} \left(\dot{a}r\delta_{\beta 0} + a\delta_{\beta 1} \right)$$

$$= \frac{a}{\sqrt{1-kr^2}} \left(\dot{a}rh^{\alpha 0}h^{01} + ah^{\alpha 0}h^{11} - \dot{a}rh^{\alpha 1}h^{00} - ah^{\alpha 1}h^{01} \right)$$

$$= \frac{a}{\sqrt{1-kr^2}} \left(ah^{\alpha 0}h^{11} - \dot{a}rh^{\alpha 1}h^{00} \right) .$$

Now,

$$K^0 = \frac{a}{\sqrt{1-kr^2}} ah^{00}h^{11} = \frac{-a}{\sqrt{1-kr^2}} \frac{\left(1-kr^2\right)a}{a^2} = -\sqrt{1-kr^2} ,$$

$$K^1 = \frac{-a}{\sqrt{1-kr^2}} \dot{a}rh^{11}h^{00} = \frac{\dot{a}ar}{\sqrt{1-kr^2}} \frac{\left(1-kr^2\right)}{a^2} = Hr\sqrt{1-kr^2} ,$$

and the components of the Kodama vector are

$$K^\mu = \left(-\sqrt{1-kr^2}, Hr\sqrt{1-kr^2}, 0, 0 \right) \quad \text{(comoving coordinates)}. \tag{A.9}$$

The norm squared of K^a is

$$K^a K_a = g_{00}(K^0)^2 + g_{11}(K^1)^2 = -(1-kr^2) + \frac{a^2}{1-kr^2} H^2 r^2 (1-kr^2)$$

$$= -\left(1 - \dot{a}^2 r^2 - kr^2 \right) \doteq -\left(1 - r^2/r_{\text{AH}}^2 \right) ; \tag{A.10}$$

it vanishes at the apparent horizon

$$r_{\text{AH}} = \frac{1}{\sqrt{\dot{a}^2 + k}} . \tag{A.11}$$

Reference

1. Nielsen, A.B., Visser, M.: Production and decay of evolving horizons. Class. Quantum Grav. **23**, 4637 (2006)

Index

© Springer International Publishing Switzerland 2015
V. Faraoni, *Cosmological and Black Hole Apparent Horizons*,
Lecture Notes in Physics 907, DOI 10.1007/978-3-319-19240-6

Printed in the United States
By Bookmasters